Tsunami

Tsunami

The World's Greatest Waves

James Goff and Walter Dudley

OXFORD
UNIVERSITY PRESS

Oxford University Press is a department of the University of Oxford. It furthers the University's objective of excellence in research, scholarship, and education by publishing worldwide. Oxford is a registered trade mark of Oxford University Press in the UK and certain other countries.

Published in the United States of America by Oxford University Press
198 Madison Avenue, New York, NY 10016, United States of America.

Library of Congress Cataloging-in-Publication Data
Names: Goff, James R. (James Rodney), 1959– author. |
Dudley, Walter C., 1945– author.
Title: Tsunami : The World's Greatest Waves /
James Goff and Walter Dudley.
Description: New York, NY : Oxford University Press, [2021] |
Includes bibliographical references and index.
Identifiers: LCCN 2020042460 (print) | LCCN 2020042461 (ebook) |
ISBN 9780197546123 (hardback) | ISBN 9780197546147 (epub) |
ISBN 9780197546130 (updf) | ISBN 9780197546154 (online)
Subjects: LCSH: Tsunamis. | Tsunamis—Research. | Natural disasters—History.
Classification: LCC GC219 .G64 2021 (print) |
LCC GC219 (ebook) | DDC 551.46/37—dc23
LC record available at https://lccn.loc.gov/2020042460
LC ebook record available at https://lccn.loc.gov/2020042461

DOI: 10.1093/oso/9780197546123.001.0001

1 3 5 7 9 8 6 4 2

Printed by Sheridan Books, Inc., United States of America

To tsunami survivors throughout the world and to the dedicated scientists who strive every day to better understand the nature of tsunamis.

Contents

Acknowledgments

A book of this nature represents decades of work with many colleagues and friends. We are grateful to so many people that it is impossible to name all of them, and we apologize for their omission. There are a few in particular who have been wonderfully supportive and provided excellent advice throughout the years.

I (JG) thank Bruce McFadgen, Darren King, Mike Archer, Chris Turney, Bruce Jaffe, Bruce Richmond, James Terry, Nigel Winspear, Kazuhisa Goto, Daisuke Sugawara, John Terrell, Gabriel Vargas, Pedro Andrade, and many more. There have been many people whom I have met on this journey who have both inspired me and helped me in so many ways. There are so many of you, but I make specific mention of a few: Kimina Lyall, who shared her experiences with us from her time in Thailand and whose book *Out of the Blue* is an insightful read; John Coney, who continues to work tirelessly behind the scenes to help make so much of our tsunami research accessible to so many people; and Robin Cain, who in recent years has provided a remarkable sounding board for all things tsunami-related.

I (WD) acknowledge the pioneering and dedicated tsunami research carried out by the following individuals, as well as the inspiration and encouragement they gave me: Doak Cox, George Curtis, Dan Walker, Frank Gonzalez, Eddie Bernard, Chuck Mader, Michael Blackford, Jim Lander, Harold Loomis, Gerard Fryer, Brian McAdoo, Lori Dengler, and Brian Atwater. I also recognize and thank the dedicated staff, board, and volunteer docents of the Pacific Tsunami Museum in Hilo, Hawaii, who tirelessly strive to educate the public about the ever-present danger from tsunamis and the appropriate preparedness and response actions to take so that those lost to deadly tsunamis may not have died in vain. And last, but definitely not least, I thank Jeanne Johnston, a tsunami survivor who fled the April 1, 1946, tsunami near Hilo and would later become a founder of the Pacific Tsunami Museum and instrumental in collecting the hundreds of survivor interviews that the museum shares with the world, excerpts of which are presented in this book.

We thank our families for all of their endless support. James is particularly indebted to Genevieve Cain for her support, love, and encouragement throughout the entire process.

Glossary

This glossary is short because we explain as we go; however, the following terms are not covered:

Epicenter (epicentre or epicentrum) The location on Earth's surface directly above the hypocenter or focus, the actual point (sometimes deep in the earth) where an earthquake or an underground explosion occurs.

Glacial moraine Any glacially formed accumulation of unconsolidated glacial debris that occurs in both currently (increasingly fewer) and formerly glaciated regions on Earth.

Kanji Chinese logographic characters used in the Japanese writing system. The Japanese term *kanji* for the Chinese characters literally means "Han characters."

Seismic waves Energy waves generated by an earthquake or other earth vibration (atomic bombs and volcanoes) that travel both within the earth (primary or "P" waves) and along its surface (secondary or "S" waves, so named because they arrive after P waves).

Seismograph An instrument used to measure and record ground motions, such as those caused by earthquakes, volcanic eruptions, and explosions.

Introduction

> What we saw that day, it's like a monster just stood up from the channels and jumped.
>
> —**Salaevalu Ulberg, Namua, Samoa**

A devastating tsunami is usually an event that is so far beyond people's life experiences that they often struggle to rationalize what they saw. The destruction that has been wrought on human communities over the millennia is reflected in stories handed down either through word of mouth or in the writings of experts and non-experts alike. It is an impossible task to do justice to all of this information, and so we have cherry-picked from the almost endless litany of disasters. We make no apology for that, but it is hoped that the reader will see a method in this madness. Examples come from throughout the world, from different sources and different times. In one sense, it is an homage to those who survived, and in another it is a glimpse into the varied world of tsunamis.

Tsunami(s)

There are many books that wisely tell people that the term *tsunami* means "harbour wave," or specifically when written in Japanese, it has the Kanjis "津" and "浪" or "波", a definition that includes any form of wave that would be unusually large inside a harbor. In one way, they are correct, and in another they are not. Tsunamis (note the Anglicized plural usage) have traveled a long way from this simple Japanese term, with its etymology having an interesting but convoluted history. It is now a globally recognized term with a far more technical definition that is

> a series of traveling waves of extremely long length and period, usually generated by disturbances associated with earthquakes occurring below or near the ocean

> floor. Volcanic eruptions, submarine landslides, and coastal rock falls can also generate tsunamis, as can a large meteorite impacting the ocean. (Goff, Terry, Chagué-Goff, and Goto, 2014)

This is an accurate, albeit somewhat dull, definition that lacks the exotic appeal of Kanjis.

Now that we have got past the scientific definition, the really important thing to think about is what tsunamis mean to people, and that is what we try to get across in this book—sometimes scientific, sometimes unusual, sometimes deeply cultural, but *always* involving people.

The reason why we have studied tsunamis for most of our careers is because there are two ways of thinking about them. The first is as a natural process that can occur on the planet. As humans living on this planet, it is important for us to know as much about all the natural processes it (and we) experiences. The second is as a hazard process capable of causing harm to people and communities and those things we value. By understanding these two processes, we strive to educate, inform, and save lives.

And so to the book.

For no apparent reason, it is probably good to outline a few bits and pieces about tsunamis or rather how we try to protect people from them because we have made a lot of progress here. True, it is never enough, but the study of tsunamis is not very old, and although that is unfortunate, we do have the benefit of modern technology at our disposal and are using it as much as we can. With that in mind, we come to the tsunami warning system.

Following the disastrous 1946 tsunami from Alaska (Chapter 1) that killed 159 people across the Hawaiian Islands, a tsunami warning system was finally established by the US Coast and Geodetic Survey in 1948 and called the Seismic Sea Wave Warning System, a name that was later changed to the Pacific Tsunami Warning System. The system was composed of seismograph observatories at College and Sitka, Alaska; Tucson, Arizona; and Honolulu, Hawaii; and tide stations at Attu, Adak, Dutch Harbor, and Sitka, Alaska; Palmyra Island; Midway Island; Johnston Atoll; and Hilo and Honolulu, Hawaii. The tide stations were to confirm that a dangerous tsunami had indeed been generated by large earthquakes recorded on the seismographs. The Honolulu (seismic) Observatory, located at Ewa Beach on the island of Oahu, was made the headquarters.

The system gave warnings for a number of tsunamis during the 1950s, none of which resulted in fatalities. That would seem to be a good thing, but some of the tsunamis were very small, causing little or no damage, and as such the public began to view these alerts as either "false alarms" or something interesting to go and watch. That all changed with the 1960 Chilean tsunami that killed as many as 6,000 people in Chile; 61 in Hilo, Hawaii; and 139 on the far side of the Pacific in Japan (Chapter 14). There had to be a way to improve the tsunami warning system to eliminate the needless warnings and evacuations and restore trust in the system. At that time, the warning system had access to numerous seismic stations throughout the Pacific, so determining the epicenter and size of an earthquake wasn't the problem. But to confirm if a tsunami had actually been generated and to determine its potential size, the system depended on measuring the tsunami waves at tide stations, ideally some close to the earthquake to provide lead time for a warning. But tide stations by their very nature, being located in harbors (i.e., protected areas for ships to load and unload), are not the best places to measure the devastating raw energy of tsunami waves at sea. The ideal place to confirm if a tsunami has been generated and just how big it is would be in the deep ocean offshore of those areas most susceptible to large tsunami-generating earthquakes.

Unfortunately, due to a lack of government funding and no devastating Pacific-wide tsunamis during the 1970s and 1980s, which might have forced the government to provide needed funding, little progress was made. Finally, in the 1990s, thanks to brilliant, innovative scientists at the Pacific Marine Environmental Laboratory outside Seattle, a prototype of such a system was developed and deployed off the coast of Oregon in 1995. It proved successful. The system is now known as DART, which stands for Deep-ocean Assessment and Reporting of Tsunamis. Basically it consists of a surface buoy and a seafloor bottom pressure recorder (BPR) shown at the lower left in Figure I.1-see color plate section, that detects pressure changes caused by tsunamis. The BPR transmits the pressure information to the surface buoy, which in turn sends the information to a satellite, which then sends the data to the National Oceanic and Atmospheric Administration's Tsunami Warning Centers.

When a large enough earthquake occurs, typically magnitude 7 or greater, seismic stations quickly determine the location of the epicenter. If the quake is near the ocean where it could potentially generate a tsunami, then the DART buoys become critical in determining if a tsunami was actually generated and how big it might be. As the tsunami waves travel across the ocean and reach the nearest DART buoy, the system takes measurements and transmits the data to the Tsunami Warning Centers, where they can make a much more accurate forecast of the tsunami and determine if a Tsunami Warning should be issued.

By 2004, the DART system had been put into operation as part of the National Data Buoy Center and had vastly improved the tsunami warning system. There would be no more "false alarms." If a Tsunami Warning is issued, that means a dangerous tsunami is on the way.

With this outline of the tsunami warning system in place, it is clear that our understanding of tsunamis has grown through time. This growth in understanding goes far deeper, however, because the warning system reflects simply one part of the story—that of the hazard. As a natural process, though, we have traced tsunamis back in time to at least 3.47 billion years ago (in Australia), and during this ongoing process of enlightenment we have learned a lot.

Couple this scientific knowledge with an unprecedented wealth of previously unpublished tsunami survivor stories available to the authors, we take you through several cases in which tsunamis have changed the world, thinking, people's lives, the way we understand these events, or all of these.

At first glance, the chapter orders may seem slightly strange. To start with, they are not in chronological order and nor do they follow a logical geographic pathway. However, there is a reason for the order in which the chapters are presented, but rather than take you step by step through the reasoning, we will just start you at the beginning.

Chapter 1 starts in 1946. In many ways, the 1946 tsunami that came out of Alaska marks the beginning of a global interest in tsunamis. It was the year that the term "tsunami" made the jump into the English-speaking world. It upset the Hawaiian Islands, and in doing so it engaged a group of US scientists in trying to unravel the details of this event. The seeds of the tsunami warning system were sown. By the time modern researchers were able to start recording survivor stories, this was the terminus ante quem. We could only trace multiple accounts of events from living survivors as far back as 1946. It marked the recognition that we really needed to know a lot more about tsunamis.

Chapter ending segues lead the reader through our logic, and we trust that this logic makes sense. However, please do not feel chained to this order. This is a book from which you can create your own order or follow ours . . . as you wish.

1
The Case of the Disappearing Lighthouse

> Scotch Cap Light was built in 1903. It consisted of a wood tower on an octagonal wood building 45 feet high and was 90 feet above the sea. . . . It was the first station established on the outside coast of Alaska. Prior to the introduction of the helicopter, access to the stations was so difficult that it was impractical to arrange for leave of absence in the ordinary way. Instead each keeper got one full year off in each 4 years of service. . . . In the 1920s and 1930s the light station underwent many improvements. 1922–23, the Navy installed radiotelephones at the station. In 1940 a new concrete reinforced lighthouse and fog-signal building was erected near the site of the original lighthouse.
>
> **—USBeacons.com (https://www.usbeacons.com/lt.cgi?lighthouse=Scotch+Cap+Light)**

It is difficult to know where to start a book of this nature. We certainly didn't want to just trot through a chronological list of events; rather, we wanted to travel around the world and look at different types of tsunamis. However, which one comes up first? In the end, it is one that was fundamental in getting tsunamis on the global map, one that saw widespread devastation, the first to truly see the gathering together of survivor stories for public education, and that introduced the word "tsunami" to the English-speaking world.

The Line Went Dead

As a tsunami researcher, it is still sometimes difficult to get the message through to people about the dangers faced from these waves. Relatively recent events such as the 2004 Indian Ocean and 2011 Japan tsunamis were brutally devastating and yet rapidly fade from people's memories as media attention moves elsewhere. In a strange quirk of fate, there are images from older events that somehow seem to have more power. Perhaps it is the distance of time or the fact that they occurred at a time when media were less fickle. This is the story of one of those images.

The Scotch Cap lighthouse lay perched on a granite cliff 100 feet (30 m) above the sea, facing the Pacific Ocean and standing as a sentinel to guide ships into the Bering Sea through the extremely treacherous Unimak Pass. It was manned by five US Coast Guardsmen, and on the bluff above it was a radio direction [direction finding (DF)] station with its own crew.

Just before 3:30 in the morning of April 1, 1946, both crews at Scotch Cap felt a strong earthquake. It was one of the largest earthquakes in US history, and compared to the magnitude 9.0 earthquake that occurred in Japan in 2011, at a magnitude of 8.6, one would think that the tsunami generated by it would have been smaller.

It was bigger.

Why? There is some debate in the tsunami community about this, no surprise there, but it may well be because it was not just a massive movement of an undersea fault. The earthquake occurred on the Aleutian Trench, which is at the contact between two tectonic plates—the Pacific and North American. The former is subducting (going under) the latter and in doing so forms a steep-sided trench up to 25,663 ft (7,822 m) deep. When this is shaken during an earthquake, it is quite possible that bits of it will fall down, with one or more huge underwater landslides adding to the problem. They too can generate tsunamis and so may well have added to the one caused by the earthquake.

However, the 1946 event is one of a special group of earthquakes known as "tsunami earthquakes"—earthquakes whose tsunamis are disproportionately larger than expected for the size of the event. This is considered to be caused by the fault rupturing extremely slowly. It is the fault rupture or the temporary "unzipping" of part of the earth's surface, as opposed to the earthquake, that generates a tsunami. Such events are generally referred to as "earthquake-generated" tsunamis. Anyway, the bottom line is that the precise mechanism that caused this huge tsunami is not entirely sorted out yet.

The earthquake was still very large, and so when the shaking finally stopped, the crew at the DF station called down to the lighthouse to check on the safety of its crew—they were all safe and there had been no damage (Figure 1.1).

A mere 48 minutes later, the US Naval Air Station at Dutch Harbor, Alaska, approximately 100 miles (160 km) southwest of Scotch Cap, just happened to be in communication with the lighthouse when they heard a Coast Guardsman scream, "A wave! Oh, no!"

The line went dead.

Figure 1.1 (a) Scotch Cap Alaska lighthouse prior to April 1, 1946. Note the DF station on the bluff above the lighthouse. (b) Scotch Cap Alaska lighthouse following the April 1, 1946, tsunami.

Source: NOAA National Geophysical Data Center (2012). Natural Hazard Images Database (Event: April 1946 Unimak Island, USA Images). NOAA National Centers for Environmental Information. doi:10.7289/V5154F01 (accessed April 20, 2020).

The wave also flooded parts of the DF station on the bluff above the lighthouse, waking the off-duty crew. They quickly struggled to higher ground, and when they looked below, they saw that there was no light from the lighthouse. In the early morning light, they could just make out the ruins of the lighthouse. There was no trace of the five-man crew, no bodies, no clothing, nothing. The following message was immediately sent from the DF station to other Coast Guard installations:

> TIDAL WAVE PRECEDED BY EARTHQUAKE COMPLETELY DESTROYED SCOTCH CAP LIGHT STATION WITH LOSS OF ALL HANDS

The killer waves of the tsunami were also heading south across the wider Pacific, where some 2,000 miles (3,200 km) away lay the isolated atoll called French Frigate Shoals, part of the northwestern Hawaiian Islands. After the Battle of Midway during World War II, an airstrip had been built on the largest of the tiny islands, Tern Island, and it was staffed by the US Navy. Tern Island is only 6 ft (~2 m) above sea level. It was now slightly more than 3 hours since the earthquake occurred and, not surprisingly, the Navy crew who were off duty and still sleeping had not felt it, but they were suddenly woken up by honking and yelling. When they ran out to see what was going on, they saw that the airstrip was almost completely covered with water. "The more we looked the more scared we got," said Petty Officer Leland Edtl. The Chief then ordered all the sailors to high ground . . . but what high ground was there? Nearly all of the buildings, like the control tower, were made of wood and were considered to be too frail to withstand the seawater. So, "the Chief ordered us all to go to the top of the ammo bunker; its top was a whopping 15 ft (4.5 m) above sea level. The highest ground on the island. There we spent the rest of a long and painful day," and they watched in fear as the water rose and fell.

Fortunately for them, low islands typically don't cause tsunami waves to rise to great heights. Such islands essentially don't get in the way of the wave, and so it passes over them as opposed to piling up against them. The biggest waves usually strike higher islands, such as the main Hawaiian Islands, toward which the tsunami was now rapidly and inexorably moving (Figure 1.2).

Some 550 miles (880 km) southeast of Tern Island lay the island of Oahu. Not too far away from the island, the destroyer–minesweeper, USS *Thompson,* was heading toward Pearl Harbor, returning from Bikini Atoll, where it had been involved in preparing for Operation Crossroads, atomic bomb tests scheduled for July. Seaman Perry Minton 17, a former resident of Honolulu, had awoken hours before dawn and was in the ship's radio room waiting for the island of Oahu to show up on the radar screen. The regular radar man had

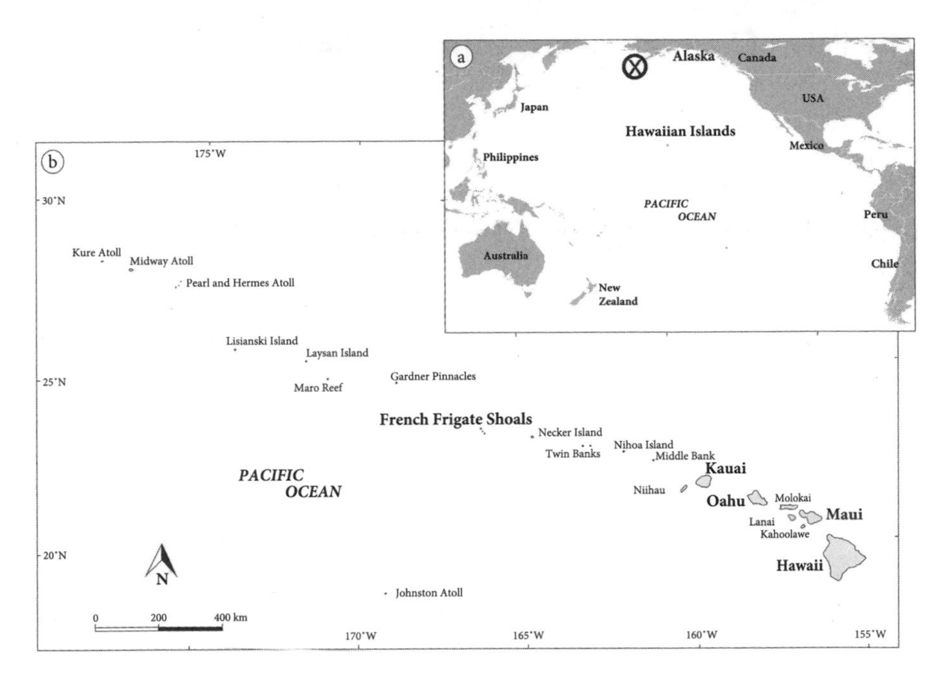

Figure 1.2 (a) The Hawaiian Islands in relation to Alaska and the rest of the Pacific. (b) Detail of the Hawaiian island chain (names in bold are mentioned in the story); the circled "x" is where the 1946 earthquake occurred that generated the tsunami. *Source*: J. Goff.

asked Perry to fill in for him while he ate breakfast. Minton said, "Almost as soon as I put on the headset, I heard a patrol plane calling its base at Kaneohe to report something on the surface of the sea, perhaps just a line or small wave." When the station at Kaneohe asked the pilot to drop down closer to the surface in order to identify the phenomenon, the pilot said it was gone: "It had outrun his aircraft." Minton wondered what was going on out there in the ocean north of Oahu, but his friends aboard the USS *Thompson* assured him that it was just an "April Fools' joke."

By this time, the tsunami had already struck the island of Kauai, where it killed 16 people along the north shore and 1 person on the southeast side of the island. Next in line was the most heavily populated of the islands, Oahu. Here, too, the northeast coast was most heavily impacted with 6 deaths, and there was just 1 casualty on the southwest shore. The island of Maui was next in line, where 14 died—13 along the northeast shore facing Alaska and 1 on the more protected western side. But the tsunami had saved the worst for the island of Hawaii, the Big Island, largest and tallest in the entire chain.

The Fate of Laupahoehoe—April 1, 1946

The small peninsula of Laupahoehoe sticks out into the Pacific Ocean along the northeastern-facing shore of the Big Island. It is the only peninsula along some 50 miles of steep cliffs until the town of Hilo farther south—a perfect target for the incoming tsunami. Once home to a sugar mill and a small village, with the coming of the railroad in the early 20th century, nearly everything had been moved up above the cliff. The only things that remained on the low peninsula of Laupahoehoe were a few small houses and the school, with seaside cottages for teachers. Idyllic.

Here there is a need for context, so let's go back a year to the spring of 1945 when young Marsue McGinnis was about to graduate from a university in Ohio with a major in Education. She intended to become a schoolteacher and like most new graduates would hope to find a job teaching at a school somewhere nearby. But this all changed one day in March when she glanced up at a billboard that stated, "Teachers desperately needed in Hawaii." This would change Marsue's life and nearly end it.

Marsue crossed the Pacific from San Francisco only days after World War II ended, but now Hawaii should be truly safe and secure. "Finally, we got to Laupahoehoe, which was just gorgeous. And we were in cottage number one. And our front yard . . . was the ocean. . . . It was just beautiful." Marsue shared the cottage with three other new teachers who had just arrived from the US mainland—Dorothy Drake, Helen Kingseed, and Fay Johnson.

The new teachers were warmly welcomed into the local community, which at that time was dominated by sugar plantations. Fall semester went well, and Marsue and her three roommates traveled to Honolulu over Christmas vacation and, according to Marsue, "had a ball."

And now it was April 1 and only a week until spring vacation, when the four were planning another trip to Honolulu for yet more fun. As Marsue recalls,

> It was about 6:45 . . . awakened by a knock on the door, "Come see the tidal wave!! Come see the tidal wave!!" . . . and being curious, we looked down and lo and behold, the water did suck out—something like emptying a bathtub. And then the water came in very slowly, like filling a bathtub. . . . That's a tidal wave? . . . So we turned around to go back in, and they said "It's doing it again!" And by that time, the kids who came to school early were watching the water suck out and trying to catch the fish. Oh, we got to take a picture of this and so I took my camera and Faye Johnson came out with me on this little porch in front of the house, good view of the ocean. And I can remember, famous last words saying, "I hope it'll be a big one this time, so I can get a really good picture.' "

> And it started to come in and it got bigger and bigger and didn't break, it just kept getting bigger. So, I dropped the camera, Faye and I tried to go down the stairs to run away and I remember turning around and seeing the, the water just fighting at the windows and then the window glass broke and the whole cottage just collapsed.
>
> And I was hanging on to the roof and so was Faye and I remember Helen Kingseed was right beside me and I put my arm under hers, but she was just swooshed right out, and I never saw her again. And so Faye and I just kinda climbed up on the roof and the roof was floating and rushing toward the school, and we got up on the cone of the roof and we were riding that and then it started to suck out again. And we clung to the cone of the roof and then the roof went clunk, and caught on the rocks. So our only hope was while it was sucking out again, to climb down off the roof. . . . And we got just part of the way and the tidal wave hit again. And so that was it. And I saw Dorothy Drake clinging on to the corner of the roof and she was just wild-eyed. . . . And we called "Dorothy come up here" and, we never saw her again, never saw Helen again. And Faye and I were kinda holding hands and making our way, and then I never saw Faye again. And I thought this is it, this is the end, I'm, I'm dead . . . and when I finally came up by this lighthouse I was surrounded by rubbish, parts of boards, you know everything plants, trees everything, coconuts. And I clung on to several boards that were still nailed together. And I thought every bone in my body is broken, but no. I could move my legs, move my arms and you know. Everything seemed to be there and working and then I noticed that my shoes were both gone. My socks were gone. My jeans were gone. My underwear was gone. The only thing left was my bra and my shirt. Thank God I had this huge man's shirt on. So I buttoned up the shirt . . . and then pretty soon along came a floating door. So I got on to that and let my other rubbish go. And then, all my thoughts were, you know "I'm the only one out here, how did I survive?"
>
> An hour went by, and I was seasick, but I knew I was going to survive, no matter what, cause I was in one piece.

As Marsue rode her door raft up an ocean swell, she saw three boys on a raft made of debris. They were too far away to yell to, but she did notice something peculiar—their faces were as white as snow. Meanwhile, Marsue continued drifting through the afternoon and she began to get more worried. Finally, she heard the sound of an airplane. It flew over and dropped a life raft. But it was too far away and she thought she could never get to it. It was getting dark and Marsue was losing hope, when the plane flew over again and "he dropped a second one, and it was very close. So I maneuvered over with my door. . . . I clambered into the little boat and let my door go." Marsue was now in a rubber boat, but it was still getting dark and she was still drifting out to sea.

Rescue

David Kailimai, the superintendent of the Hamakua Mill Company, had been at his home in a village a few miles up the road from Laupahoehoe when the waves struck. About an hour later, one of his workers called him to say that there had been a "big flood" and trucks were needed to salvage possessions. When he drove down the steep road to Laupahoehoe and saw the floating debris and the leveled buildings, he knew that there had been a tsunami.

He knew also that a boat was essential to save people, about 10 of whom he could see floating in the ocean. No help could be expected from Hilo because it seemed likely to David that the need for boats in Hilo would be equally great and that many vessels would have been damaged or destroyed. His own boat was on the other side of the island, in Kona, but he did have an outboard motor at his house. The only boat in the Laupahoehoe area was a sailboat that belonged to Mr. Walsh. At first, Walsh was not willing for his boat to be used because the sea was still rough and full of debris, the wind was strong, and it seemed likely that his boat would be damaged. Moreover, the boat would have to be cut to accommodate David's motor. It took much persuasion, and by the time the boat was launched it was after 2:00 p.m. The action of the wind and waves had dispersed the people David had seen earlier. He set off to search for survivors accompanied by the local physician, Libert Fernandez. Dr. Fernandez was anxious for the safety of his fiancée, a young teacher. First they found two boys, students from the school; then a seaplane, flying overhead, directed them to where it had dropped a rubber raft. On the raft they found Marsue McGinnis, the fiancée of Dr. Fernandez. In Marsue's words, "Dr. Fernandez's boat came out. And there I was saved."

Herbert Nishimoto

Herbert Nishimoto would also never forget what happened to him on that April Fools' Day. A 10th grader, he lived in the tiny sugar plantation village of Ninole, where his parents owned a general store. He was a Boy Scout and a good swimmer. He practiced swimming in a local stream, using rocks to channel the water and form his own "endless pool."

There had been a sophomore class picnic at the school on Saturday, so Herbert and a couple of buddies decided to spend the rest of the weekend in an empty teachers' cottage next to the ocean on Laupahoehoe Point. They awoke to the sound of a friend, Daniel Akiona, running past the cottages shouting "Tidal wave!" and he urged them to go to his family's home on the other side of the point.

Herbert watched the waves come and go, and he saw the second wave demolish an old canoe shed by the shore. But the third wave took longer to gather than the others and, according to Herbert, "looked terrifyingly huge." Desperate for refuge, they dashed toward Daniel's home, but as they reached the door the whole building collapsed and Herbert was lifted by the wave, sucked into the ocean, and deposited on the rocks along the shore. As he gasped for breath and stared around him, he saw his friend Mamoru floating on a log and a teacher floundering in the water. Herbert tried to remove his jeans so that he could swim more easily, but they were so tight around his ankles that he had only just managed to get out of one trouser leg when the next wave arrived. As he dived under the wave, the heavy trousers trailing from one foot caught on the rocks and he was battered to and fro. When the force of the wave finally subsided, he found himself floating in among the debris and accompanied by sharks! Fortunately, his luck continued. A section of flooring from one of the cottages floated nearby, and he heaved himself onto it and lay wondering what would happen next. "I found a mattress and some logs and I made a raft."

Herbert was now stranded in the ocean on his improvised raft, but after a while he saw two fellow students, Takemoto and Kuhuki, and managed to pull them aboard. Not long after that, he saw a bottle of Crisco shortening for cooking floating by and grabbed it so that they could rub it all over their bodies to protect them from the sun and cool wind and water. They were indeed the three strange, white-faced boys that Marsue saw that day, and they were also the target of the first life raft Marsue had seen dropped from the airplane.

The boys managed to reach the raft and quickly climbed into it, rowing away from the debris fearing that nails or splinters in the wood might puncture the little rubber craft. Darkness fell, and between spells of dozing off, they tried to determine where they were along the coast. They finally saw a light on the cliffs above, and Herbert realized that they had drifted northwest all the way to the town of Honokaa, some 20 miles from Laupahoehoe. The sun came up and Herbert recognized Waipio Valley, where there was a beach. However, they were too far offshore to make it there, and they continued to drift past other deep, uninhabited valleys. If they couldn't get to shore soon, they would end up in the notoriously dangerous Alenuihaha Channel, which in Hawaiian means "great billows smashing."

Ahead were mostly the steep cliffs making up the shoreline of the Kohala sugar plantation, but there were a few tiny bays that didn't look very welcoming. And then,

> I seen a plane circling . . . we waved, but they can't see us. So I had a mirror in the rubber raft made out of stainless steel, so I shined that light against the airplane. I did that for maybe 5, 10 minutes and he finally seen us so he came circling around us and that's when I seen a kid running on the hill. And that was just about 11 'o clock.

The kid on the hill was a little girl heading to her grandmother's house who had spotted the boys in the raft and alerted plantation workers who were going home for lunch. Herbert recalled,

> And then I seen a whole bunch of people on the beach there running down you know from the hill running to the shoreline area. Two guys hit the water and start swimming out. So we start oaring in, and as soon as we oared in, the rubber raft turned over. So I jumped on it, I grabbed, I was trying to grab the other 2 guys but they were gone. I couldn't find them then the 2 swimmers says "I got one," then the other guy finally got the other one. Then they took us on shore.

They all survived, but sadly, 15-year-old Daniel Akiona, the classmate who first warned Herbert, was killed by the tsunami. His name is one of 24 on the monument at Laupahoehoe Point, in remembrance of those who died there that fateful day (Figure 1.3).

Hawaiian Afterword

Following my (WD) interview with Herbert Nishimoto, he mentioned that he had saved the oars of the life raft that had been dropped to him following the tsunami and that they were in his garage. He donated them to the Pacific Tsunami Museum in Hilo, Hawaii, where they are on display, along with his story. When I asked Marsue McGinnis what advice she would give to people to help them be better prepared for the next tsunami, she said, "Whenever I've told my story, I always try to emphasize that it's not a wave, it's a series of waves. It goes out, comes in, the next one is bigger. Evacuate, get to higher ground right away or you're going to be swept away."

Excellent advice from somebody who miraculously survived this powerful tsunami.

Figure 1.3 Monument at Laupahoehoe Point to those lost to the April 1, 1946, tsunami. *Source*: Walter Dudley Collection, Pacific Tsunami Museum.

But the Tsunami Moved On . . .

The tsunami had caused chaos in the Hawaiian islands, but it moved on. In the Marquesas Islands of French Polynesia, some 2,800 miles (4,300 km) to the south, the waves reached even higher to 65 ft (20 m) but killed only two people. Chile's Easter Island (4,600 miles or 7,500 km southeast of Hawaii) experienced waves nearly 28 ft (9 m) high, and in the Juan Fernandez Islands (another 1,800 miles/2,900 km east) they were approximately 9 ft (3 m) high. Pitcairn Island, New Zealand, Samoa, and California were all struck by waves. One person was killed in California and another in Peru, 7,000 miles (11,000 km) away from where it all began.

And so onwards . . . to Antarctica.

The western coast of the Antarctic Peninsula is more than 9,300 miles (15,000 km; the equivalent of flying from New York City to the very southern tip of New Zealand) from the source of the tsunami. On the morning of April 2, 1946, 21 hours after the earthquake, the tsunami struck.

Wordie House is located on Winter Island, one of the Argentine Islands in the Wilhelm Archipelago of Antarctica. It is a classic Antarctic research hut with a kitchen, living room, toilet, and just enough creature comforts to make the long work days survivable. It was established on January 7, 1947—after the tsunami—and is named after James Wordie, the chief scientist and geologist on Shackleton's *Endurance* expedition of 1914–1917. However, the hut stands on the foundations of an earlier building, used by the British Graham Land Expedition in 1935 and 1936. The original hut was destroyed in 1946 by the tsunami, which was at least 13 ft (4 m) high. The hut floated off its foundations, and debris was scattered along the coastline of the adjacent Skua Island.

Wow.

The widespread destruction, from the disappearing lighthouse to the shattered Antarctic hut, appears to mark a tsunami watershed in both Japan and the English-speaking world. It was at about this time that the term *tsunami* became the one to use when talking about these waves. Although it had been used in Japan for many years before, it had competed with other terms such as *kaishou* (海嘯). *Tsunami* (津波) became the prominent term between 1937 and 1946, when the Japanese Education Ministry adopted the story of *Inamura no hi* as a textbook for part of the elementary school syllabus. This story describes how Gohei saved his villagers from the 1854 Ansei Nankai tsunami (discussed in Chapter 2). In this textbook, the term *tsunami* (津波) was used, and *kaishou* (海嘯) did not appear. During this period, the 1944 Tonankai and 1946 Nankai tsunamis occurred in Japan, and many young people described these as a "tsunami" because they had learned this term and how to pronounce it from the textbook. In the English-speaking world, the term was "borrowed" by some remarkably astute US scientists (MacDonald, Shepard, and Cox), who not only used it in discussions of the effects of the April 1, 1946, event in Hawaii but also explained what it meant. The term appears to have had an easy transition in Western science, becoming almost immediately globally accepted by the scientific community.

2

How Weird Squiggles Led from Sheaves of Rice to the Depth of the Seas

"It is not normal," Gohei muttered to himself as he came out of his house. The earthquake was not particularly violent. But the long and slow tremor and the rumbling of the earth were not of the kind old Gohei had ever experienced. It was ominous. Worriedly he looked down from his garden at the village below. Villagers were so absorbed in the preparation of a harvest festival that they seemed not to notice the earthquake. Turning his eyes now to the sea, Gohei was transfixed at the sight. Waves were moving back to the sea against the wind. At the next moment the expanse of the sand and black base of rocks came into view.

"My God! It must be the tsunami," Gohei thought. If he didn't do something, the lives of four hundred villagers would be swallowed along with the village. He could not lose even a minute.

"That's it!" he cried and ran into the house. Gohei immediately ran out of the house again with a big pine torch. There were piles of rice sheaves lying there ready for collection. "It is a shame I have to burn them, but with this I can save the lives of the villagers." Gohei suddenly lighted one of the rice sheaves (Figure 2.1). A flame rose instantly fanned by the wind. He ran frantically among the sheaves to light them. . . .

The fire of the rice sheaves rose high in the sky. Someone saw the fire and began to ring the bell of the mountain temple. "Fire! It is the squire's house!" Young men of the village shouted and ran hurriedly to the hill. Old people, women and children followed the young men. . . . The villagers gathered one by one. He counted the old and young men and women as they came. The people looked at the burning sheaves and Gohei in turn. . . .

At that time he shouted with all his might. "Look over there! It is coming." They looked through the dim light of dusk to where Gohei pointed. At the edge of the sea in the distance they saw a thin dark line. As they watched, it became wider and thicker, rapidly surging forward.

Figure 2.1 Drawing of Goryo Hamaguchi (Gohei), a village leader in Hirogawa (formerly Hiro village), Wakayama Prefecture, who set fire to piled sheaves of his newly harvested rice in order to warn people about the tsunami on December 24, 1854.
Source: The Cabinet Office of Japan.

"It is the tsunami!" someone cried. No sooner than they saw the water in front of them as high as a cliff, crashing against the land, they felt the weight as if a mountain was crushing them. They heard a noise as if a hundred thunders roared all at once. The people involuntarily jumped back. They could not see for a while anything but clouds of spray which had advanced to the hill like clouds.

They saw the white fearful sea passing violently over their village. The water moved to and fro over the village two or three times. On the hill there was no voice for a while. The villagers were gazing down in blank dismay at the place where their village had been. It was now gone without a trace, excavated by the waves.

The fire of the rice sheaves began to rise again fanned by the wind. It illuminated the darkened surroundings. The villagers recovered their senses for the first time and realized they had been saved by this fire. In silence they knelt down before Gohei.

—Account from Japanese village of Hiro, Wakayama Prefecture, Japan (after Tsuchiya and Shuto, 1995)

Deep Thought

The remarkable events just described occurred on December 24, 1854. This was the Ansei Nankai earthquake and tsunami. The earthquake was a magnitude 8.4, and the resultant tsunami ran up onto the land as high as 36 ft (11.0 m). This was a terrible time for people living on the east coast of Japan because the Ansei Nankai earthquake, along with the tsunami that followed, was the second of three Ansei great earthquakes. The first one, a day earlier—the Ansei Tōkai earthquake and tsunami—was equally devastating. The third one, the 1855 Ansei Edo earthquake, did not generate a tsunami. However, immediately following this event, prints started to be produced showing a giant catfish known as Namazu-e. Its thrashing around under the surface of the earth was commonly considered to be the cause of earthquakes. However, the vast numbers of prints produced at this time reflected a period of significant political and social unrest following these disasters and the recent appearance of the Black Ships of Commodore Perry in 1853. The ships were so named in reference to their color and the color of the smoke produced from the coal-fired steam engines. This was gunboat diplomacy, and Commodore Perry's show of military force was being used as a key player in negotiating a treaty allowing American trade with Japan.

Times they were a-changing.

The Ansei Tōkai tsunami occurred 150 years and 3 days prior to the disastrous Indian Ocean tsunami of December 26, 2004. There would be no official warning for either tsunami, but what transpired in the United States following Gohei's heroic sacrifice in Japan could and should have changed all that.

Alexander Bache, the head of the US Coast Survey, had overseen the installation of the first tide gauges along the US west coast. In July 1854, just 6 months prior to the earthquake and tsunami in Japan, a tide gauge was installed at Fort Point on the grounds of the Presidio (today underneath San Francisco's Golden Gate Bridge). This gauge would go on to record the longest continuous series of tidal observations anywhere in the Western Hemisphere.

In December of that year, the tsunami waves from Japan had crossed the Pacific Ocean and had been recorded on the new tide gauge as strange squiggles, very different from the lines created by the normal rise and fall of the tides. The bizarre squiggles on the tide gauge were noticed by Lieutenant Trowbridge, who monitored the tide gauge at Fort Point, and he wrote to his boss: "There is every reason to presume that the effect was caused by a submarine earthquake." That was a remarkable insight considering that recording seismographs would not be invented for another quarter of a century. Bache, however, was skeptical and wrote,

> In February, 1855, I received a letter from the Pacific coast calling attention to the self-registering tide-gauge at [San Francisco and] San Diego, on the 23rd . . . of December [1854], and remarking that the irregularities of the curve could not be produced by disturbances from storms, as the meteorological records for the whole coast showed a continuance at that time of an ordinary state of weather. . . . There is no reason to presume that the effect was caused by a sub-marine earthquake. . . . No shock . . . has been felt at San Francisco.

As it happened, Captain Adams of the US Navy had witnessed the tsunami firsthand at Shimoda, Japan. He mentioned the tsunami in a letter to Commodore Perry, who had recently visited Shimoda for the ratification of a treaty between the United States and Japan.

This was an interesting treaty. The Kanagawa Treaty, as it was known, was signed in March 1854 and ratified on February 21, 1855. It was effectively signed under threat of force (the Black Ships) and meant the end of Japan's 220-year-old policy of national seclusion, opening up the ports of Shimoda and Hakodate to American vessels. A Russian delegation trying to do the same thing for their ships arrived in Japan in November 1854 to continue negotiations that began again on December 22. The next day, a 23-foot (7-m)-high tsunami destroyed most of Shimoda, including all the ships of the Russian delegation. Stranded in Japan, the highly civilized Russian representative was able to negotiate the long-awaited Russo-Japanese treaty of friendship on February 7, 1855. More generous terms were granted to the Russians than to both the Americans and the British largely because the main Russian negotiator viewed his Japanese colleague as a father figure and deferred to him as such. He made a good impression.

A mere 2 weeks later, Captain Adams, who had been in Japan for a while, was in Shimoda representing the United States as the two countries ratified the less generous treaty, which Japan signed under duress.

On the tsunami, Adams wrote,

> The sea rose in a wave five fathoms [30 ft, ~9 m] above its usual height, overflowing the town and carrying houses and temples before it in its retreat. . . . Only 16 houses were left standing in the whole place. The entire coast of Japan seems to have suffered by this calamity.

As fate would have it, a chance encounter in Washington, DC, on June 20 of that year between Commodore Perry and Bache would change everything.

Perry told Bache about the tsunami in Japan. Bache would now see the bizarre squiggles in a different light. Interestingly, he was the great-grandson of Benjamin Franklin and must have inherited some of his great-granddad's scientific curiosity. Bache analyzed the records of the tide gauges at both San Diego and San Francisco and found that they had indeed recorded small waves from the tsunami. Based on accounts of the tsunami, it had occurred at approximately 9:00 a.m. in Japan. The squiggles on the tide gauges in California began approximately 13 hours later. Knowing the time of the earthquake and the time of the tide recordings, he calculated the speed of the tsunami waves from Japan to California. Using this speed and the recently published Airy wave theory (by Sir George Biddell Airy in his treatise on waves and tides, *Encyclopaedia Metropolitana*, 1849), Bache was able to calculate for the first time the average depth of the North Pacific Ocean. His estimate of 15,000 ft (~4,500 m) for the track from Shimoda to San Diego is remarkably close to the modern estimate of 15,221 ft (4,640 m). Previous estimates (by French scientist Pierre Simon de Laplace) suggested depths as great as 60,000 ft (~18,000 m). The speed of tsunami waves had now been successfully used to calculate the depths of the seas, and Bache published his findings in 1855.

Years later, we would have measured the depths of the seas in many places. Could that knowledge of ocean depth be used to calculate wave speed and then to accurately predict the time when tsunami waves would come crashing ashore? First, we needed some advance warning that a tsunami might have been generated. Would knowing that earthquakes could cause tsunamis and the detection of seismic waves be used to predict a potential tsunami?

An Early Effort at Tsunami Warnings: The Jaggar Story

In 1912, the Hawaiian Volcano Observatory (HVO) was established, and on the staff were a number of scientists who began to study earthquakes and tsunamis. Thomas A. Jaggar, the founder of HVO and director until 1940, began to investigate and report on tsunamis and even researched historical accounts of earlier tsunami events. Jaggar knew that the seismic waves caused by earthquakes are transmitted across the world in a matter of minutes. A large earthquake in Chile, for example, would be registered on seismographs in Hawaii hours before a tsunami could reach the islands. Why not use this lead time to warn of an impending tsunami?

In early 1923, Jaggar had the opportunity to witness the earthquake–tsunami relationship firsthand. As he inspected the seismograph at 8:00 a.m.

on the morning of February 23, he noticed the trace of a large earthquake that had been recorded earlier that morning at 5:32 a.m. Hawaiian time. Jaggar quickly calculated that the epicenter would be approximately 2,500 miles (~4,000 km) away, possibly under the sea off the Aleutian Islands. The seismic waves had taken only approximately 7 minutes to travel through the Earth to Hawaii from the Aleutians; if tsunami waves were on their way, they would arrive several hours later. Jaggar notified the Harbor Master in the town of Hilo of the possibility of a tsunami later that day, but his warning was not taken seriously.

At 12:30 p.m., the tsunami struck Hilo, almost exactly 7 hours after the earthquake had hit the Aleutians. The largest wave was the third of the series, rising to more than 20 ft (6 m) and carrying the local fishing fleet of sampans from their moorage in the river into and under a railroad bridge. Most of the sampans were smashed to bits, and one fisherman was decapitated when his boat was forced under the bridge by the power of the tsunami waves. It was a painful lesson, but it did highlight the possibility of protecting life and property by warning of approaching tsunami waves.

Later that same year, meteorologist R. H. Finch gave a speech at a scientific meeting in Sydney, Australia. The title of his talk was "On the Prediction of Tidal Waves" (published the next year in the *Monthly Weather Review*), a subject that as Jaggar prophetically stated, "might well be studied to advantage." Finch noted that the time in hours it took tsunami waves to reach a site was approximately equal to the time in minutes it took the seismic waves to arrive there. Using the minutes (for earthquake waves) = hours (for tsunami waves) rule, he believed that it should be possible to predict the arrival time of tsunami waves from all areas of the Pacific. Finch suggested also that because most seismographs were inspected rather infrequently, some type of "alarm bell" should be attached to the instruments to alert scientists when a large earthquake was registered.

Jaggar believed that the study of tsunamis was now more important than ever. He urged that to accurately predict the time it took the waves to travel from their sources to Hawaii, it would be necessary to correlate actual earthquakes with the arrival times of tsunami waves. He also thought that by more intensively studying earthquakes and tsunamis, it would be possible to determine the earthquake intensity required to generate a tsunami. Fortunately, most tsunamis are very small and go unnoticed unless registered on a tide gauge. In order to benefit from the study of these small tsunamis, Jaggar traveled to Washington, DC, to arrange with the US Coast and Geodetic Survey for the establishment of a tide gauge for Hilo Bay, not far from HVO.

With tide gauges now in Honolulu and Hilo, it was possible to examine the records of changes in water level from these gauges and identify even small tsunami waves of 1.5 ft (0.5 m) or less. With seismographs, tide gauges, and interested and knowledgeable scientists in Hawaii, the connection between earthquakes and tsunamis became better known. In 1933 came another opportunity to test the earthquake–tsunami relationship. At 7:10 a.m. Hawaiian time on March 2, a large earthquake was registered on the seismographs at HVO and the distance to the epicenter calculated to be 3,950 miles (~6,400 km), possibly off the coast of Japan. Knowing that a tsunami might possibly have been generated, the Hilo Harbor Master was contacted and told that waves might begin to arrive at approximately 3:30 p.m. that afternoon. Would he listen this time? Fortunately, remembering the 1923 tsunami, the Hilo sampan fleet was moved out to anchorages in the bay.

At approximately noon in Hawaii, radio news broadcasts announced that a disastrous earthquake had occurred in Japan. In fact, the earthquake (magnitude 8.6) and accompanying tsunami in Japan resulted in more than 3,000 lives being lost and nearly 9,000 homes and some 8,000 boats destroyed. The tsunami waves were highest along the northeast Sanriku coast of the main Japanese island. Here, waves up to 75 ft (23 m) high were reported at Hirota Atumari. Tsunami waves were also spreading across the Pacific at nearly 500 miles (800 km) per hour.

On the west side of the island of Hawaii (the Big Island), the side facing Japan, the first waves of the tsunami arrived at 3:20 p.m. local time. At first, the sea withdrew, exposing wide areas of the sea floor in bays; then canoes and other small craft were torn from their moorings and capsized. As the sea returned, walls were knocked down, and houses were flooded and washed away. Of the series of some 10 waves, the last was the most damaging, with a total vertical range of 17.5 ft (5.3 m). On the east side of the island in Hilo, the waves began to arrive at 3:36 p.m., only 6 minutes later than predicted, but the water only rose and fell a total of 3 ft and caused no property damage.

As a result of the warning, there was no loss of life in Hawaii. But warnings of tsunamis based on earthquakes alone could also lead to false alarms, and government officials believed that a tsunami warning system based solely on the occurrence of submarine earthquakes was virtually useless.

Even by 1946, many scientific experts had yet to fully understand tsunami waves. Dr. Francis Sheppard, a distinguished marine geologist with Scripps Institute of Oceanography, was staying with his wife in a beach-side cottage on the north shore of Oahu Island in Hawaii and provided the following account of the April 1, 1946, Aleutian tsunami:

> We were sleeping peacefully when we were awakened by a loud hissing sound, which sounded for all the world as if dozens of locomotives were blowing off steam directly outside the house. Puzzled, we jumped up and rushed to the front window. Where there had been a beach previously, we saw nothing but boiling water, which was sweeping over the ten-foot top of the beach ridge and coming directly at the house.

After witnessing the arrival of the first wave, Sheppard got his camera, left the house, and saw the water retreating until the coral reef was exposed and stranded fish were flapping.

> Trying to show my erudition, I said to my wife. "There will be another wave, but it won't be as exciting as the one that awakened us." . . . Was I mistaken? In a few minutes as I stood at the edge of the beach ridge in front of the house, I could see the water beginning to rise and swell up around the outer edges of the exposed reef; it built higher and higher and then came racing forward with amazing velocity. . . . As it piled up in front of me, I began to wonder whether this wave was really going to be smaller than the preceding one. I called to my wife to run to the back of the house for protection, but she had already started, and I followed her just in time. As I looked back I saw the water surging over the spot where I had been standing a moment before. Suddenly we heard a terrible smashing of glass at the front of the house. The refrigerator passed us on the left side moving upright out into the cane field. On the right came a wall of water sweeping down the road. We were startled to see that there was nothing but kindling wood left of what had been the nearby house to the east.

As the wave subsided, the Sheppards hurried to higher ground, just ahead of a third and still larger wave.

> We started running along the emerging beach ridge in the only direction in which we could get to the slightly elevated main road. . . . As we hurried through this break, another huge wave came rolling in over the reef and broke with shuddering force against the small escarpment at the top of the beach. Then, rising as a monstrous wall of water, it swept on after us, flattening the cane field with a terrifying sound. We reached the comparative safety of the elevated road just ahead of the wave. . . . Finally, after about six waves had moved in, each one apparently getting progressively weaker, I decided I had better go back and see what I could rescue from what was left of the house. . . . I had just reached the door when I became conscious that a very powerful mass of water was bearing down on the place. . . . I rushed to a nearby tree and climbed it as fast as possible

and then hung on for dear life as I swayed back and forth under the impact of the wave.

The Sheppards both survived the 1946 Aleutian tsunami unharmed, and Dr. Sheppard learned valuable lessons that would soon appear in a scientific research paper. However, immediately following the 1946 tsunami, the Commander of the Coast and Geodetic Survey stated the government's case: "Less than one in one hundred earthquakes result in tidal waves and you don't alert every port in the Pacific each time a quake occurs."

An uphill struggle to save lives lay ahead.

Despite government claims that no warning of the 1946 tsunami had been possible, it was obvious that something had to be done to protect the population of Hawaii. Both civilian and military sources criticized the US Coast and Geodetic Survey for not issuing a warning. After all, as critics noted, the seismic waves from the earthquake had been recorded at HVO within minutes after the earthquake struck the Aleutians; consequently, a tsunami could have been predicted.

An official tsunami warning system was finally established by the US Coast and Geodetic Survey in 1948 and was initially called the Seismic Sea Wave Warning System; it was later renamed the Pacific Tsunami Warning System. The system was composed of the US Coast and Geodetic Survey seismograph observatories at College and Sitka, Alaska; Tucson, Arizona; and Honolulu, Hawaii; and tide stations at Attu, Adak, Dutch Harbor, and Sitka, Alaska; Palmyra Island; Midway Island; Johnston Atoll; and Hilo and Honolulu, Hawaii. The Honolulu (seismic) Observatory, located at Ewa Beach on the island of Oahu, was made the headquarters.

Initially, the warning system was to supply tsunami warning information to the civil authorities of the Hawaiian Islands and to the various military headquarters in Hawaii for dissemination throughout the Pacific to military bases and to the islands in the US Trust Territory of the Pacific. But even as the warning system was being set up, memories of the devastation of 1946 were beginning to fade, at least in the minds of the government agencies in Washington, DC, that were responsible for funding the system. In fact, during its early years under the Department of Commerce, the funding was practically nonexistent. An anecdote recounted by Bernard Zetler, a former manager of the warning system, illustrates the lengths to which the scientists had to go to secure funding in the early 1950s. A colleague of Zetler, Harris B. Stewart, happened to meet the Congressional Delegate from Hawaii at a social function and explained the importance of the warning system. The delegate offered to write to the Secretary of Commerce suggesting increased support for tsunami warnings if Stewart would draft the letter for him. When the Secretary of Commerce received the letter from the senator, he sent it to Stewart and Zetler and asked them to draft a reply. An exchange of several letters followed, with neither the delegate nor the secretary realizing that Stewart and Zetler were drafting both ends of the correspondence. In the end, the funding was increased slightly, but the warning center continued to be funded on a shoestring budget compared with most other government agencies.

We have come a long way since then. Indeed, we have now arrived at a point where the need to better understand tsunamis is treated seriously. It is a science in its own right, and it calls upon a vast array of data and information in order to ultimately protect lives (Figure 2.2-see color plate section). Here, we have seen how the ability to identify tsunamis through the instrumental record has slowly developed over time—this is great. But in explaining this journey, we have drawn not only upon personal accounts from 1854 and 1946 but also upon the story of Gohei and the sheaves of rice. The account of Gohei from the Japanese village of Hiro would have undoubtedly been passed on initially by word of mouth, only to be written down later by a literate scribe. But these stories exist everywhere; they exist in prehistory—before the written record. It is these stories or warnings handed down through the generations or preserved in names in the landscape that warned people of such hazards. These voices from the past are returning today to guide the modern tsunami scientist.

3

Voices from the Past

> "Get back!" he cried. "Get back to the high ground, or you will be drowned," and running past his people he climbed the high cliff, where he took his stand, and repeated more spells. The people, thoroughly terrified, followed helter-skelter, and left Titipa alone upon the beach. Soon the sea grew dark and troubled and angry, and presently a great wave, which gathered strength as it came, swept towards the shore. It advanced over the sandy beach, sweeping Titipa and all his fish before it, till with the noise of thunder it struck the cliff on which the people stood. . . . The great wave receded, sucking with it innumerable boulders and the helpless, struggling Titipa. Then another wave, greater than the previous one, came with tremendous force and, sweeping the shore, struck the cliff with a thunderous roar. This was followed by a third which, when it receded, left the beach scoured and bare. Titipa and all his fish had disappeared."
>
> **—Grace (1907, pp. 158–159)**

History Isn't Always Written

A tsunami geologist has to be something of a Jack of all trades and master of none. There are so many different skill sets you need to have in order to really understand what you are looking at in the sediment when you are standing in a hole in the ground or somewhere similar. There is an old adage, or perhaps it is not that old, that a tsunami deposit (the sediment left behind after a tsunami has inundated the land) is a "deposit out of place." In other words, it is something that you don't expect to find where you are finding it. On a good day when the gods are smiling and the planets align, this may well be the case, but in reality it is always a bit more complicated than that. And this is where other trades come to help the geologist.

It may seem an obvious statement, but the reason why we worry about tsunamis is because they kill people. This has been one of the main driving reasons behind the work that both of us do—as scientists we want to learn as much about these deadly waves as possible so that we can better inform the wider community in order to reduce fatalities in the future. To understand them better, we therefore need to look for all the possible clues we can, and that is where the past comes in. We look at past events because they guide our understanding of what will happen in the future. And yes, this means we look at the geological record, but it also means we look at prehistory. Catastrophic tsunamis such as the 2004 Indian Ocean event have occurred many times in the past, and like in 2004, they have unfortunately killed many people. We can therefore find evidence for tsunamis in archaeology, a sort of geological *Time Team* (the UK name; it was called *Time Team America* in the United States) where both geology and archaeology help you unravel the story. We can also find evidence in anthropology—the study of human societies and cultures, not just the physical remains of the past—and in particular, we can find evidence through oral traditions.

Many of the survivor stories from the 2004 Indian Ocean tsunami have now been written down and recorded in some way, shape, or form and are currently being used in a wide range of educational material (discussed further in Chapter 15). There is a huge and invaluable archive maintained by the Pacific Tsunami Museum in Hilo, Hawaii. This is a veritable treasure trove of information and a place to be visited. For those of us who live in countries with a long written history, it often comes as a surprise to realize that there are still many areas of the world where everything was passed on by word of mouth until very recently, sometimes only a couple of hundred years or so ago.

This applies to numerous places in and around the Pacific Ocean. Oral traditions in the Pacific still have immense importance, and in many ways they are the "history" of an island or a country in the region. In contrast, in the British Isles, for example, we have grown up reading about our past. We were taught about it in history lessons, we have been told about old manuscripts such as the *Magna Carta* and the *Domesday Book*, and some of us may even have seen these documents. We sometimes even quibble about some of the details reported in these written records, which invariably occurs when either a new piece of evidence is found in recently discovered ancient documents or old ones are reinterpreted. But in general, we accept that if something is written down, it pretty much happened. In England, we know that a major event such as the War of the Roses took place on and off between approximately 1455 and 1487. Why do we know this? It is all written down. As a result, we tend to be a little uncomfortable around oral histories and use words such as myth, fantasy, or folklore to express this uncertainty—words that bring with them a hint of make-believe or even disbelief.

This is probably not entirely surprising, but if we dig deeper into our past, we find that many of the things we know today may well have their origins in oral tradition. At some point, this knowledge eventually made it into print, and over time these origins have been all but lost. A simple example is the River Trent in the Midlands of England. Most people have no idea where the name came from, but it is one of the few surviving British names in an area that was governed by Danelaw from approximately the 9th century. It probably survived because it was a convenient descriptive term likely meaning "great wanderer," so named because it would frequently "wander" and flood its banks, inundating the land to devastating effect. This is a meaning that is just as apt today as it must have been at the time it was named. In the United States, there are many similar names that sit there in the landscape and tell us something about the past. Sequim is a city in Clallam County, Washington, located in a bay on the edge of the Juan de Fuca Strait. It is in the homeland of the S'Klallam Indian tribe, and the bay was originally called Such-e-kwai-ing, which means "quiet water"; it was later anglicized into Sequim. Although this is undoubtedly a rich fishing area, the straits are prone to massive storms (and the odd tsunami) emanating from the Pacific, and so a quiet protected bay was, and still is, a welcome haven.

Similar oral traditions exist or have existed throughout the Pacific region, the only difference being that few have been written down, and those that have been written down may have been misinterpreted. However, with regard to oral traditions about tsunamis, invariably the descriptive nature of the language closely resembles that of accounts told to us by modern-day survivors. Could they therefore represent a direct record of people's experience with tsunamis before Europeans arrived in the region? In many areas of the Pacific, Europeans—and writing—did not arrive until the late 18th century, so the possible depth of history retained in these oral traditions could be huge.

The Taniwha

New Zealand Māori talk of mythical creatures called *taniwha.* Taniwha are often associated with dangerous places—rapids in a river, a fast-flowing channel, or a dangerous obstacle—a warning of a place to avoid or the need to be careful when crossing a river, for example. In this instance, the taniwha are essentially a metaphor for danger in the water, and if they appear here to be malevolent, it is because the stories are usually about events that killed people or caused damage. Taniwha typically come in the guise of water monsters or giant lizards that cause mayhem of one sort or another and kill and eat people.

Figure 3.1 Detail of an ancient rock drawing of a "taniwha" consuming a human figure from the top of a limestone cave at Weka Pass Range near Waikari, New Zealand (the original figures were painted red).
Source: von Haast (1877).

There is a *pūrākau* or Māori oral tradition related to the *iwi* (tribe) that lived at a place called Moawhitu (Greville Harbour) on D'Urville Island off the northwest coast of the South Island of New Zealand. The common interpretation of this tradition is that sometime around the 15th century, nearly everyone who lived there was killed by a tsunami that, as the tradition tells, was in the form of a taniwha called *Tapu-arero-utuutu* (Figure 3.1). The wave was *utu* (vengeance) because a woman had broken a food *tapu* (taboo). Tapu-arero-utuutu drowned the people, piling their bodies up in the sand dunes. In the frenzy of its attack, a piece of its tail broke off and still remains today as a rock sticking out of the water in Greville Harbour as a reminder of this event. Recent work has found geological evidence not only for this event but also for earlier ones—this is not a safe place to live. But although the tsunami was massive, the tradition is silent about whether the wave struck other communities on the island or nearby mainland. Like all Māori oral traditions, however, this has many more layers to it than simply referencing a tsunami, although it is impossible to do justice to all of them here. Could this have just been a local event or are there more serious concerns here for New Zealand?

The "Coming of the Sands"

Another Māori oral tradition is the story of Potiki-roa, a Taranaki chief who married the daughter of Mango-huruhuru, a powerful *tohunga* (priest) who lived in the South Island. Potiki-roa lived at a place called Potiki-taua, between Waitara and Cape Egmont, on the west coast of the North Island of New Zealand. On his return there after some years away, he took both his wife and his father-in-law with him. According to the tradition, there were few beaches in Taranaki suitable for hauling out canoes in those days because of the lack of sand (which is still the case today as well). To rectify this problem, Mango-huruhuru offered a *karakia* (prayer) to bring sand from Hawaiki (in

Māori oral tradition, this is the original home of the Polynesians). At the conclusion of the karakia, there was a "great storm." The sea rose and the sands came, killing people and deeply burying them, houses, gardens, and all the surrounding countryside. Potiki-roa and his wife survived because they had built their house far enough inland to avoid the sand and waves, but Mangohuruhuru, whose house was closer to the sea, perished. Again, this happened sometime around the 15th century.

Is this a coincidence? Perhaps this was just a really bad storm? What about the Māori oral tradition of the taniwha called *Te Kaiwhakaruaki* or the one named *Rapahoa* just down the coast and the giant "storm" conjured up by a wizard to form the sand dunes farther south? Not convinced? That is the problem. Those who are used to written records and modern science may roll their eyes at such stories, which is a shame because these are invaluable guides for us. Think about it: The River Trent in England sits there as a name telling us that it floods, so why do these far more explicit stories seem unreal? They are not unreal to Māori—such oral traditions are part of their history. But just to prove how powerful these oral traditions are, scientists have now demonstrated there was a huge tsunami that affected much of the central western shores of New Zealand sometime between approximately 1470 and 1510. How do we know? We have the geological evidence that matches the stories—but the stories came first, and the stories guided the modern science.

Fifteenth Century in the Southwest Pacific

New Zealand is not alone in oral traditions. After all, the Pacific Ocean is a large piece of the planet—more than one-third of it. There are 22 Pacific Island countries mostly consisting of a huge number of islands dispersed throughout the region. It is in the southwest area of this region that the tsunami story gets bigger—much bigger.

In approximately 1452 or 1453, geological evidence suggests that there was a massive eruption and the island of Kuwae, in the middle of Vanuatu (an archipelago of approximately 80 islands) some 870 miles (1,400 km) north of New Zealand, disappeared. It also left behind a huge hole in the seafloor. To put this in context, Krakatoa (the correct spelling is Krakatau; see Chapter 11), the Indonesian volcano that erupted in 1883, generated a tsunami that was well over 30 m high that killed thousands of people. Kuwae was approximately 10 times bigger! And with the eruption came an oral tradition, and also . . . a tsunami. According to tradition, the eruption was the result of man, Pae, seeking revenge on the people of Kuwae for tricking him into sleeping with his mother.

However, as if this was not bad enough, 1,100 miles (1,800 km) east of Vanuatu, the island of Pukapuka in the Cook Islands was also struck by a tsunami at approximately the same time. Only 2 women and 15 men survived, and it is from this depleted group that much of the current population of some 500 people are descended.

Are these two linked? Well, in a way yes and in a way no. This was a very bad time to be in the southwest Pacific (Figure 3.2).

Using a mixture of oral traditions, archaeology, and geology, scientists have managed to piece together the records of two tsunamis that, if they didn't occur at the same time, probably occurred within a few years of each other. The

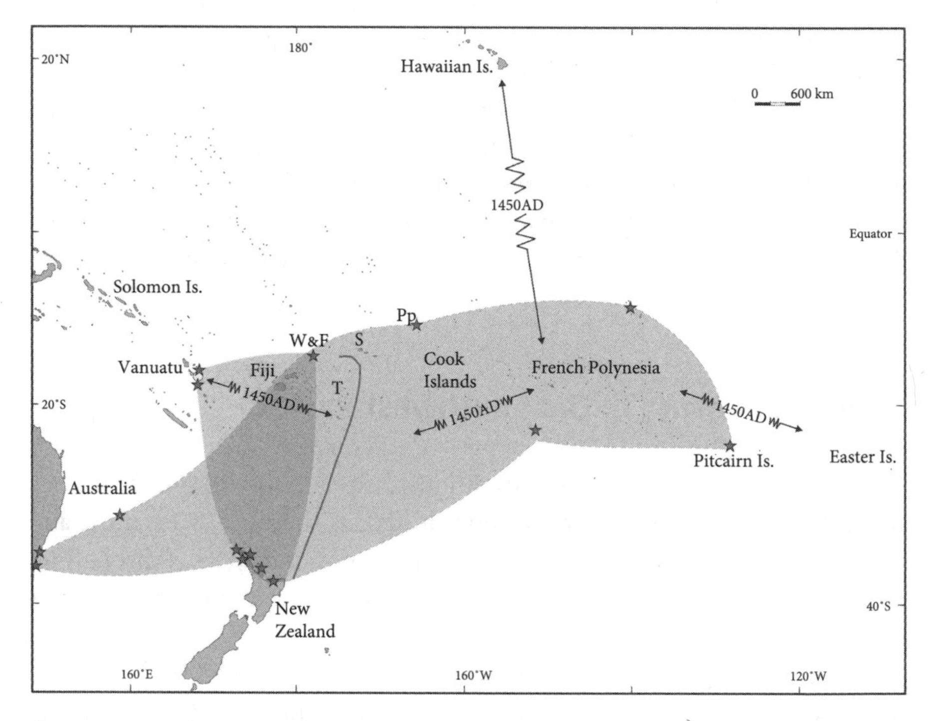

Figure 3.2 Southwest Pacific—estimated extent of two mid-15th-century palaeotsunamis. Kuwae event: Triangular shading including Vanuatu, Wallis and Futuna and New Zealand; Tonga trench event: Shading spanning area from Australia to Pitcairn Is; approximate extent of Tonga (Kermadec) trench marked by thin solid line extending from northeast New Zealand to between Tonga and Samoa; Pp, Pukapuka; S, Samoa; T, Tonga; W&F, Wallis and Futuna. The black squiggly lines indicate the breakdown in contact between widespread areas of Polynesia in the mid-15th century.
Source: J. Goff.

tsunami caused by the Kuwae eruption pales in comparison to the one generated by the Tonga trench—a huge fault line to the north–northeast of New Zealand that marks the boundary between two tectonic plates, the Australian and the Pacific. Both these events (Kuwae and Tonga trench) happened only approximately 150 years after all of Polynesia had finally been settled and at a time when there was a fantastic connectedness between the islands in the Pacific with huge long-distance voyages taking place. Trade goods from New Zealand can be found in Hawaii (that's approximately 4,400 miles or 7,000 km—try doing that and you realize the enormity of what the Polynesians achieved way before Europeans tentatively limped into the region), Tonga traded with Vanuatu, and so on.

And then suddenly—it all changed.

No more long-distance canoe journeys, islands became isolated, internal warfare and tribal bickering started, and these predominantly coastal people moved inland and uphill away from the coast (a movement that was only reversed when missionaries arrived and lured the people back down to the coast where they could be better managed—forget the fact that this then exposed island populations to tsunamis and cyclones all over again—we never learn). Oral traditions of taniwha and the wrath of the gods and other catastrophic stories were handed down from generation to generation to remember the events of this time. There was loss of the elders, most probably drowned by the tsunami(s)—these were the keepers of much of the tribal knowledge and with their loss came the loss of skills to make high-quality tools, for example—and there was a huge loss of voyaging canoes, coastal settlements, crops and other resources to the giant tsunami.

Armageddon had struck, and the entire Polynesian culture never fully recovered before Europeans discovered it. The rest, if you excuse the pun, is history.

Where Tsunamis Began?

One of the interesting things about this whole catastrophic scenario is that it took the oral traditions to guide modern science to find geological and archaeological evidence for these major events in the Pacific. We weren't really looking before, or if we were, it was most definitely not in the right places and not for evidence of such events. Such natural disasters did not make it a particularly pleasant time to be living in the Pacific, but this was followed a few centuries later by rapid European colonization, with missionary zeal forcing religion upon the peoples, largely forbidding the ongoing use of oral traditions.

It is lucky for modern science, however, that some stories have survived to guide researchers. In countries less affected by the European tidal wave, many oral traditions have survived and, not surprisingly, their (and our) understanding of tsunamis is far better as a result. For example, on the island of Hokkaido in northern Japan, there is an old Ainu oral tradition called *Umi ni ukabu yama o oyoide hippatta otasut-jin no hanashi* ("The story of Otasut-lad who pulled an island"). Here, a boy is one day visited by his uncle, who takes him on a trading trip by canoe. Just to be on the safe side, the old woman with whom the boy lives gives him a charm, a belt. The boy and his uncle depart on their trip, but after a while the uncle leaves him on an island so he can go off and trade by himself. The boy becomes angry and passes his belt around the mountainous island and pulls it toward himself. This generates a large tsunami that overturns his uncle's boat and kills him. This oral tradition is as yet undated, but there have been numerous large tsunamis during the past few thousand years off the island's eastern coast along the Kuril trench, a large subduction zone that marks the boundary between the Pacific and North American plates. But let us not forget the 2011 earthquake and tsunami that occurred to the south of here that killed nearly 20,000 people.

It is hardly surprising that Japan has many oral traditions. It is, after all, where the word *tsunami* (roughly translated from Japanese as an unusually large wave inside a harbor) originated. This is an easy word to drop into a sentence, but actually it is a rather complicated word to explain.

Prior to the Edo Period (1603–1868) and for much of it as well, there was no specific word "tsunami." In general, most people at this time were illiterate and so the details of any coastal disasters such as this were passed by word of mouth as stories to the literate few, the scribes who worked in the capital city. At this time and until 1868, the capital was in Kyoto, after which the emperor moved it to Tokyo (interestingly, the name Tokyo is a simple reversing of the two syllables of Kyoto). Passing on these stories by word of mouth meant that there was an almost endless list of phrases or words used to describe the disasters affecting different areas of the coast because almost every coastal community had different ways of explaining what happened and used different borrowed Chinese "words" to describe it. For example, one of the stories transcribed in an old (720) document called the *Nihon shoki* describes the 684 Hakuho–Nankai tsunami at Kochi as "大潮高騰, 海水飄蕩, 由是運調船多放失焉" or "large wave run high up to land and seawater inundated. Due to this, many ships with tributes were lost away." A more recent example is of the 869 Jōgan tsunami (the previous big one before the 2011 event in Japan) described as

“海口哮吼, 聲似 雷霆, 驚濤涌潮, 泝洄漲長, 忽至城下, 去海數十百里” or “sea mouth barking, the sound was like thunder. Turbulent large wave and the welled-up seawater run the river and quickly reached to the castle town that is located far away from the coast.” Neither of these somewhat lengthy phrases really captures what we can visualize today when the word “tsunami” is used.

Actually, the term tsunami (津浪), although probably spoken alongside others at approximately the time of the Jōgan event, was a rather late addition to the known written record of Japan, first appearing in 1524 with reference to the 1454 Kyoutoku tsunami. The ongoing journey of the term tsunami through Japanese language and culture provides us with an example of cultural progression from descriptive oral traditions to written records as the society became increasingly more literate. However, just because more people could read and write did not mean that other words were not used; it had great competition from the word *Kaishou* (海嘯; a Chinese word used to describe the movement of water along a river caused by tidal effects—a tidal bore—the most famous of which is probably the one in China’s Qiantang River). Part of the reason the word *Kaishou* started to become less popular was probably because in Japanese the meaning wasn’t quite right and it was also more difficult to use in written and oral communication.

A major change occurred with the 1896 Meiji–Sanriku tsunami. Japanese newspapers, now being read by an increasing number of people, generally used the word tsunami to describe the event. Interestingly, an 1896 description of this event was written in *National Geographic Magazine*—the first time the word tsunami was used in English. By the time Japan was hit by the 1944 Tonankai and 1946 Nankai tsunamis, the word was starting to be used in Japanese textbooks and as a result was generally becoming more widely adopted.

By this point, readers may be wondering why we have used italics for the word *Kaishou* but not for the word tsunami. Ultimately, this all comes down to the 1946 Aleutian tsunami that affected areas of the western coast of the United States, Alaska, and the Hawaiian Islands. In reporting on the effects of the 1946 Aleutian tsunami in the Hawaiian Islands, three remarkably erudite scientists—Doak Cox, Gordon Macdonald, and Francis Shepard—adopted the word and explained its meaning. Western scientists rapidly picked up the word and it has since become anglicized, hence no need for the italics. So, while our beloved media continued to use the incredibly unhelpful (scientifically at least) term “tidal wave,” both Western and Japanese scientists were on the ball. It took until the 2004 Indian Ocean tsunami, when the world’s media were given some serious lessons in terminology by scientists during a raft of interviews, for the supertanker of journalistic prose to start to shift.

Encouragingly, after the 2010 Chilean tsunami struck, more than 95% of all media reports were using the word tsunami—it just goes to show the power of the spoken word.

Delving deeper into the story of the word tsunami, it is interesting that while the 1896 Meiji–Sanriku tsunami saw the term really start to take off in Japan, it was also the event that saw the appearance of another word—*Yoda*. This seems to be a term used to describe a gradual rise in water level associated with a particular type of tsunami, so the phrase would be something like "look out, *yoda* is coming," no need to watch out for a light sabre. So once again, there is another Japanese word out there that gives us an even greater vocabulary to help us describe these events. It would be very useful if this and similar words were adopted from Japanese to help us better describe what we see—in the case of *yoda*, this may allow some of us to drift into *Star Wars* humor and perhaps use the phrase "Yoda strikes back." [I (JG) used it for the title of a recent paper!]

Now that we have established the term tsunami and the value of oral traditions, we have to remember that the only reason these came into being was because these events killed people, and these events needed to be remembered and recorded in some way, shape, or form. However, it is easy enough to say that prehistoric events killed people, and indeed the oral traditions tell us so, but is there actually any physical evidence that people died? After all, as noted at the beginning of this chapter, a tsunami geologist has to be something of a Jack of all trades and master of none. One of those trades is delving into the anthropological record of oral traditions, but another is archaeology—the physical evidence of past human activity in and around the coast. Did past tsunamis truly wreak the same havoc as that seen in recent events?

4
The World's Oldest Tsunami Victim at the Gateway to the Pacific—and Beyond

> While it may never be possible to definitively assign the Aitape Skull as the earliest tsunami victim in the world, this reassessment of the existing work indicates that the Sissano Lagoon region may well contain an extensive Holocene record of human interactions with catastrophic events such as tsunamis.
>
> —**Goff et al. (2017)**

We Forget

As discussed in Chapter 3, something verging on a Pacific Armageddon struck the Polynesians in the 15th century, but although to many people that might seem like a long time ago, it hardly even registers as the blink of an eye in geological terms. It should come as no surprise, then, that geologists have traced tsunamis back into a much deeper past than a mere 500 years or so. A later chapter takes us back into the millions of years, but for now it seems like a good idea to take the tsunami story back a few thousand years and continue this Pacific odyssey.

There are many places in the world that we simply do not know exist until something happens that requires people to go there. Often as these news stories unfold, we don't even know exactly where these places are or even have any grasp of the magnitude of the event unfolding there. The 2019–2020 bushfires in Australia are a classic example. It is one thing to watch the horror of these fires engulfing everything in their path, but to truly understand the enormity of this event, we need some type of reality check—they covered an area more than five times that of the state of New Jersey, approximately the size of Virginia, or approximately five times the size of Wales and nearing that of England. To say that they were catastrophic almost feels like an understatement. However, in the true tradition of quick news, short attention spans, and sensationalist media, they will rapidly fade from most people's memories when the next crisis is reported—enter coronavirus disease 2019 (COVID-19). It is amazing how few people now remember the 2004 Indian Ocean tsunami; it killed approximately a quarter of a million people, with Sweden approximately 5,400 miles (~8,750 km) away from the action suffering the country's worst loss of life from a natural disaster in recent history with 543 dead. Most were vacationing in Thailand.

How can such events be forgotten so easily?

This preamble is relevant in some way to every chapter in this book. We forget. We forget the tragedies; we move on.

In doing so, the work of the scientists so desperately sought after by the media for their opinions and knowledge in the immediately aftermath is forgotten as well. The media move on. They grab everything they need immediately from the scientists, taking up their valuable time and energy in pursuit of a decent video clip or sound bite, and then drop them like some discarded rubbish to move on and hassle the next group of experts about the next media sensation. The media are something of a double-edged sword: They get the story they want (often "massaging" the message they were given in order to get more readers), and we get . . . exposure? A chance to educate people, get our message across even if that message was not exactly what we said? Hey ho—can't live with them, can't live without them!

Once the locust swarm of the media moves on, the science—our findings—largely dwells in the somewhat obscure world of academic journals and there it remains, with the majority of it becoming increasingly old and forgotten even by the academic community. And so we come to Sissano Lagoon.

Sissano Lagoon

Sissano Lagoon sits along the remote northern coast of Papua New Guinea (PNG) on the western edge of the vast Pacific Ocean. It came to fame albeit briefly in recent years when it was struck by a tsunami in 1998. The tsunami killed well over 2,000 people—a massive number for a remote and poorly populated area of the world. The cause of this tsunami was a source of some significant scientific debate for several years. At first, it was thought to have been caused by an earthquake (the cause reported by the media), and then it was thought to have been caused by an underwater landslide (some media reported this story). Ultimately, however, these two mechanisms were, quite logically, tied together. The earthquake was really too small (magnitude 7.1) to have generated the size of tsunami that struck the coast [30–50 ft (10–15 m) high], but it did set off a landslide and the after-effects of both of these mechanisms did the deed.

Just by way of a reality check at this moment, the tsunami was therefore higher than a three-story building. We should not have been surprised by this event, even though we were. It was the third to have affected the area in 100 years, but that spans several generations and with each passing generation it gives just enough time between events for them to be largely forgotten or at best retained in traditional stories that are increasingly ignored by younger generations that have not had firsthand experience of such an event.

The fact that there have been repeat tsunamis along this coast in historic time points to what has been an ongoing hazard for thousands of years. We know this is the case and can now take ourselves back approximately 6,000 years to a time when people were just starting to move into the area. We even have skeletal evidence for them in what is known as the Aitape Skull, named after the largest town in the Sissano Lagoon area. It was found and reported in one of those old, forgotten scientific reports in 1929. The site is approximately 7 miles (12 km) inland from Sissano Lagoon and 170 ft (52 m) above sea level in sediments that were distinctly marine in origin—odd.

However, before jumping to conclusions and getting ahead of ourselves, this does not mean that an ancient tsunami more than 50 m high devastated the coastline at this time. Perhaps more remarkably, what this 52-m-high deposit does show is that these marine sediments have been lifted up to this height over approximately 6,000 years by repeated earthquakes. This averages out to only approximately 0.4 inches (1 cm) each year, which may not sound like much but it cannot be stopped. It continues in fits and starts even today; each earthquake jolts the land up just a little bit more.

Having found the skull in 1929, the researchers, who were working for an oil company and so were somewhat distracted at the time, finally got around to looking at it in more detail 20 years later in 1949. Today, their initial findings would have been splashed over the news, albeit most likely only for a day or so, because it was thought to be old—very old—as in more than 1 million years old. This news caused something of a stir in the archaeological community because it suggested that our ancestors had moved out of Africa and as far east as PNG extremely early on in human history. Such excitement was short-lived, however.

The radiocarbon dating method was developed during the late 1930s and the 1940s primarily by Willard Libby, who received the Nobel Prize in Chemistry for his work in 1960. By then, there were many laboratories throughout the world and dating human bones using this technique was well underway. The basic premise is that when an animal or plant dies, it stops exchanging carbon with its environment, and from that point onward the amount of radiocarbon it contains starts to decay at a measurable rate. So when an old bone is discovered, the amount of radiocarbon remaining can be used to work out when the owner of the bone died. In 1964, the skull was radiocarbon dated and found to be only approximately 6,000 year old. With that news, the skull fitted the time frame of when humans had started to arrive in the area, and it was essentially assigned to a dusty cardboard box in an Australian museum.

After the 1998 tsunami occurred, however, the vaguest memories of this skull and its brief moment of fame meant that it once more came under the scientific spotlight, back from the stacks of forgotten literature. It was not necessarily the skull that initially caused the interest but, rather, it was the sediment that it was found in that was curious. On re-examination, this marine sediment turned out to be a tsunami deposit—the evidence of a tsunami that hit the coast 6,000 years ago. The Aitape Skull gained new fame as probably the oldest tsunami victim in the world.

Why is this important? It means that we now know that tsunamis have been affecting human populations in this area for many thousands of years, and that we never learn. However, this also provides us with a jumping off point into the Pacific.

Slightly to the east of Sissano Lagoon, around the Bismarck Archipelago, there had been a gradual filtering down of people from Southeast Asia approximately 6,000 years ago. These people were great seafarers and navigators. They made pottery, and as opposed to being a "people" they are recognized as a culture—the Lapita culture. Again, there is huge debate not only regarding where they came from but also concerning how quickly—a rush or a trickle—but the latter is gaining favor so we stick with that.

Approximately 3,500 years ago, give or take a few hundred years, members of the Lapita culture ventured into the western Pacific. Today, in the age of planes and GPS navigation, this may not seem like a major feat because people regularly go island-hopping—the dream of many gap-year students to this day. But the amazing thing was that they could not see the next island to hop to, and there were no GPS, maps in the form we know today, or airplanes or speedboats to make that hop. These were impressive navigators who could sail against the wind or at the very least had an intimate knowledge of the wind that meant that if they couldn't sail against it, they knew the wind direction was going to change and so they weren't too phased about heading off in one direction because they knew they could get back. Their knowledge of the sea was fantastic. Even today, we are still learning about the intricacies of how they navigated.

Here, two examples are presented. First, they employed stick charts made of coconut fronds that were tied together to form an open framework, with island locations represented by shells tied to it. There were threads that represented prevailing ocean waves and their directions as they changed when they approached islands and met other similar wave crests. Each chart was created by a particular navigator—their crib sheet for their mind map of the long journey across the Pacific Ocean. There was (and is) also testicular navigation, in which the navigator would sit cross-legged on the bottom of the canoe and feel the nature of the ocean swells through an extremely sensitive part of his body. These swells are consistent across thousands of miles of ocean, and if one can read the shape of a swell, one can determine the direction and strength of the current beneath it. To understand this means that one can steer a canoe in the right direction and not drift off course into the empty expanse of the ocean.

So yes, one needs balls to navigate the Pacific!

Why was there this trickle of migration into the western Pacific from PNG? Good question. Take your pick—resource depletion, warfare, natural disasters, all of these, or something else. The truth is that we are not entirely sure, but the point is that people moved, and quickly. They arrived in Tonga, Fiji, and Samoa at approximately the same time some 3,000 years ago. It was a 2,200-mile (3,500-km) journey of settlement over 500 years across an awful lot of ocean littered with islands, many of which were permanently settled, ultimately forming what archaeologists now call western Polynesia. However, Polynesians didn't officially become Polynesian until the ability to make Lapita pottery died out some 2,500 years ago. It is a somewhat random choice for a name change, but it will do.

How do we know that the Lapita culture spread though this area? There have been numerous archaeological digs. Pottery, old settlements, and even rare burials have been found. In addition, quite recently, archaeologists working in Vanuatu [~1,200 miles (2,000 km) away from PNG] found the earliest dated mass burial Lapita site—more than 25 skeletons buried higgledy-piggledy in gaps in an old coastal reef. Most of these burials can be dated to approximately 3,000 years ago, which fits well with our current understanding of the spread of people across the western Pacific. We can never expect to find the first people who did this, but this recent find is undoubtedly early on in the story, analogous to getting close to the Big Bang.

And here we come again to an interesting aside, or rather a logical question that we should ask at this point: Why did these people in Vanuatu die? Why the mass burial in what most definitely seems to be a rather hasty manner with bodies essentially squeezed in among the coral? There are of course many possible explanations, but most of the obvious ones can be discounted. For example, warfare—no, there are no major fighting injuries. Famine—nothing obvious. Ritualistic burial—no appeasing of the gods here, although there was some of what is termed "funerary practices" that generally represent what we perform today. As we rapidly tick off the list of usual suspects, we have to start looking around for unusual suspects; after all, we seem to be dealing with a mass burial here, so it was not just the normal attrition of everyday life.

And here we have, entering from left field, a tsunami. Could this mass burial of some of the earliest settlers in Vanuatu be related to deaths caused by a tsunami, just like PNG and the Aitape Skull?

It had to happen at some point, and indeed that is rather the case with tsunamis. If only we could predict them, but tsunamis just happen. They are unlike cyclones, which, as a seafaring people, these people would have known were going to hit days in advance. However, it is fair to say that tsunamis are mostly generated by earthquakes, which could be used as a warning sign of impending doom. But this is not by any means the only way they are generated, and they can often occur without warning or without time to respond. This was in an age in which the people were on their own; there was no Pacific Tsunami Warning System and no email, internet, telephones, or radio—just the vast expanse of the Pacific Ocean.

Geologists working in New Zealand and the southwest Pacific have found evidence for a tsunami that affected much of the region approximately 2,800–3,000 years ago, all the way from Australia to New Zealand and Vanuatu and undoubtedly beyond. Such an event would have been devastating for a seafaring and coastal culture. Canoes would have been destroyed, villages and crops obliterated, and intertidal shellfish species ripped up by the waves and killed. Many people would have been killed, caught largely unawares by the waves. Survivors would have struggled to get on with their lives, largely ditching convention and rapidly burying bodies close to where they died to possibly deal with them later in a more appropriate manner. Life would have been tough. Resources depleted and wary of the sea, they most likely moved away from the coast to metaphorically lick their wounds. With death came the loss of much knowledge handed down by oral traditions, and as such their ability to bounce back would have been hampered, leading to a time of regrouping and relearning the skills needed to be seafarers again.

This may sound a little far-fetched, but let us not forget the devastation caused by the 2004 Indian Ocean tsunami and the 2011 Japan one as well: Without massive international aid and engineering, humanitarian, and economic efforts, the coastal communities affected by those events would no longer exist. So if this tsunami actually did happen, there must be some evidence left in the culture other than a mass burial (on that point, there were 14 mass burials alone in Banda Aceh, Indonesia, following the 2004 tsunami; the largest contained 60,000–70,000 people).

What is perhaps the most compelling evidence is the start of what archaeologists call "the long pause." Archaeologists acknowledge that the Lapita culture moved rapidly eastward across the Pacific and most definitely arrived in the Tonga–Samoa archipelago by approximately 2,800 years ago. Given the distances involved, the island-hopping and settling in at each place—remember that as far as we know, there was not any real drive to keep going east (in modern terms it was probably what we would call an organic process; it just happened)—this was a remarkably fast settlement of the western Pacific and then . . . nothing. Here around Tonga–Samoa they stopped, and they stopped for a long time, not going any farther east until approximately 1,000 years ago, a nearly 2,000-year "long pause," with just a minor little wobble approximately 2,000 years ago when a few adjacent islands seem to have been settled.

So the major question has always been—Why?

Taken out of context, we can think of many reasons:(1) They got bored of traveling; (2) the supply chain backing up this big move had to catch up with the rapid advance (similar to the push by the US Army in World War II when General George Patton had to wait for fuel and supplies to catch up with his rapidly advancing tanks and troops); (3) there was enough room for everybody in this archipelago so why go any farther; or (4) beyond this point, they ran out of useful rock to use for tools—no more andesite equals no more movement. The andesite line separates the mafic basaltic volcanic rocks of the Central Pacific Basin (material generated by "gentle" eruptions such as those experienced in Hawaii) from the partially submerged continental areas of more felsic andesitic volcanic rock on its margins (explosive stuff), and this line sits immediately adjacent to the Tonga–Samoa archipelago on its eastern flank. This is a convenient archaeological explanation but one that really doesn't seem to fit the bill because when they did move, they seemed to adapt well to the alternative resources available to them.

Let us add a little more geological context. The andesite line on the eastern flank of the Tonga–Samoa archipelago exists because of the Tonga–Kermadec trench, or Tonga Trench, a very long and large subduction zone akin to the ones that wrought havoc through the earthquakes and tsunamis of 2004 and 2011 around Indonesia and Japan, respectively. If the trench here is like the other ones that have been so brutally active during the past decade or so, then it also undoubtedly has the ability to generate similarly devastating tsunamis. Is this what stopped the easterly movement of the Lapita culture? In other words, did these skilled seafarers face a similar tsunami some 2,800–3,000 years ago as those that occurred recently?

Well, we now have a mass burial in Vanuatu that fits with timing of such an event and geologists have also found evidence for a massive tsunami throughout the region at that time—a perfect reason for a long pause. Let us remember that this seafaring culture would have been well versed in the vagaries of waves and storms and cyclones, however severe. They lived in and understood their environment far better than we do today. We have air-conditioning to keep us cool and heating to keep us warm and to buffer us from changes in the weather and obscure the environmental signals that tell us something bad is coming. In the absence of a weather report, we are at the mercy of the elements, especially in these remote Pacific islands. The Lapita culture undoubtedly had an intimate knowledge of meteorological indicators that gave them advanced warning of such events, but earthquakes and tsunamis are a different beast. Yes, they may well have experienced them occasionally in their travels, but in reality it was only upon entering the wider Pacific that they were exposing themselves to the full might of the double punch of giant earthquake and massive tsunami, and the event approximately 3,000 years ago would have been their first true welcome to the Pacific Ring of Fire. If we infer that they were indeed caught unawares, then it is no wonder that there was a long pause, and this is probably where oral traditions of such events started, although that is pure speculation because we can really only trace oral traditions of earthquakes and tsunamis back to more recent events a few hundred years ago.

Did anything happen during the long pause? Well, not a lot. The Lapita culture seems to have lost its ability to produce Lapita pottery, and this seems to be a good point to define the beginning of Polynesians approximately 2,500 years ago, although there seems to be one exception to this rule. People on the island of Futuna, part of France's territory in the Pacific—the Wallis and Futuna archipelago—seemed to hang on to this skill until approximately 2,000 years ago. This is important because of what *did* happen during the long pause. As briefly mentioned, approximately 2,000 years ago there were minor settlement expansions in western Polynesia into islands such as Rotuma, Pukapuka, and Niue. Why? Well, that's a good question. Also, how synchronous was it? The answer to both these question is that we are not sure. However, what we do know is that there was another reasonably large tsunami around this time that affected the region, and the key evidence for this was found on Futuna (Figure 4.1). It inundated the island at a time when there was settlement along the

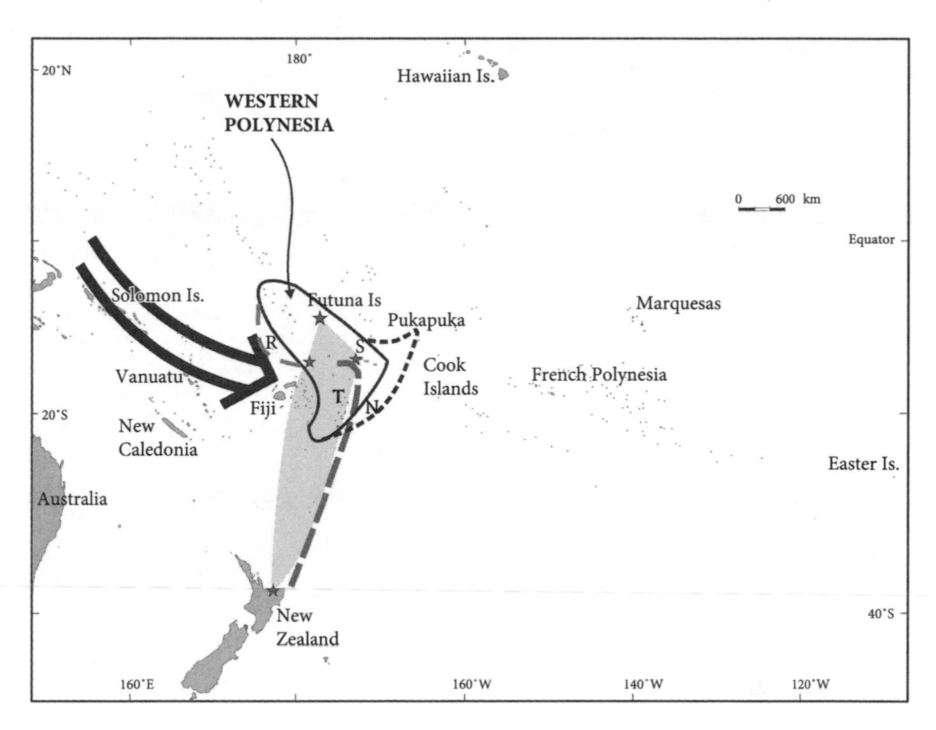

Figure 4.1 Western Polynesia: People migrated from the west (thick black arrow) and stopped here approximately 3,000 years ago, and then 2,000 years ago a minor shuffling around of people in these islands occurred (dashed black line) at the same time as a large tsunami. The grey shaded area shows all that is currently known of the extent of this event. N, Niue; R, Rotuma; S, Samoa; T, Tonga. Approximate extent of the Tonga trench is marked by a thick dashed grey line.
Source: J. Goff.

coastline, and within the deposit there was a lot debris from this settlement, including some Lapita pottery—evidence that pottery making was still being practiced here and indicating perhaps that this event was its death knell because there is no evidence of pottery making after this tsunami. It ultimately died out in this last refugia approximately 2000 years ago. On Futuna, this 2,000-year-old tsunami is marked by archaeological evidence for settlement movement inland and uphill to escape these unpredictable and devastating events (Figure 4.2). However, in the true tradition of these things, people had resettled the coast of Futuna by the time the next large tsunami occurred approximately 500 years ago. Movement inland and uphill following a tsunami is a common trait, as is the reoccupation of coastal sites in the quiescent phase that follows. Humans do not have a good track record of doing what is best for them—later missionary zeal didn't help either—and this same pattern has

Figure 4.2 Futuna: The 2,000-year-old event (1) overlies an early occupation layer. A later tsunami in the 15th century (2) overlies a subsequent occupation once people had moved back down to the coast from the hills.
Source: J. Goff.

occurred over the millennia in most countries affected by tsunamis, with both Indonesia and Japan being prime examples in which recent disasters have reaffirmed the stupidity of humans.

And so, the long pause continued until slightly less than 1,000 years ago, and then suddenly the settlement of eastern Polynesia began, picking up Easter Island by approximately 750 years ago. Let's put that last statement in context: That is approximately 4,000 miles (6,500 km) in 250 years compared to 2,200 miles (3,500 km) in what seemed like a pretty impressive effort in the first stage between 3,500 and 3,000 years ago. Not only that, though, Easter Island is not exactly close to any other islands. The Polynesians probably took a circuitous route to get there via French Polynesia—the nearest islands being some 1,600 miles (2,600 km) to 2,000 miles (3,200 km) away. In other words, one canoe trip, no "stopovers" to rest weary arms, and let's not forget the navigational skills required to get there. It was not until 1722 that Easter Island was found by Europeans, and it was almost another 50 years before anyone else turned up. These Polynesians were brilliant seafarers. Approximately 100 years after the end of the long pause, they had visited South America some 2,200 miles (3,500 km) farther east, where they introduced some chickens to

Chile and obtained sweet potatoes in return; they had settled Hawaii 4,500 miles (7,300 km) from Easter Island; and they finally arrived in New Zealand (they would have known about the place, probably even visited it, but let's be fair—compared to all those lovely tropical islands, it is a bit chilly) to stay around 1300, give or take a few years. Oh yes, that's approximately 4,400 miles (7,100 km) away. This is not to say that they headed off to all of these places from Easter Island, but it does demonstrate the kind of massive long-distance trading routes that existed and essentially kept many of these isolated islands in business. Not all the resources they needed were available, so trade was very important. Not only did they find and settle these islands but also they survived through this incredibly sophisticated long-distance trade voyaging throughout the Pacific.

And then—*nothing*. It stopped, people moved inland and uphill, resources became depleted, warfare increased, fort building increased, and their stone tool technology took a backward step of several millennia.

What happened? Have you read Chapter 3, "Voices from the Past," yet?

If you have, then you know; if not, the 15th century was not a good time to be in the South Pacific. Two events (that we know of)—a massive volcanic eruption and yet another massive earthquake—both generated huge tsunamis that effectively demolished coastal Polynesian settlements. They were caught out again by the surprise attack; these were not predictable meteorological events. The geological evidence may not look like much, but to geologists it is easy to read and every bit as impressive as the 2011 Japan tsunami.

These catastrophic events were sufficiently recent for oral traditions to have survived, and we discussed those in Chapter 3. However, let us briefly think about today. Earth scientists still debate how big and how often events such as these can happen in the region and yet the Polynesian legacy recorded in the archaeology and anthropology is sufficient to tell us that these things happened. However, it does not tell us that they will happen again, and although the debates will continue doubtless for many years to come, we as a society are even more unaware of our environment.

Who do you know who could jump into a canoe and navigate without GPS or any modern trappings all the way to Easter Island? How do we know when a storm is coming? We cannot do even these simple tasks. How on earth can we even think that we are truly prepared for the next tsunami?

When the next big earthquake occurs and you are on a Pacific Island or having a great time on a beach somewhere, what do you do? Maybe of course it will not be generated by an earthquake; it could be a landslide, which is much more of a silent killer, requiring an even more intimate knowledge of the environment for you to spot it. Many may not remember, but there are many photographs of people standing on the beach just staring at the 2004 Indian Ocean tsunami as it moved rapidly and inexorably toward the shore. Why did people just stand there and look at it? Well, quite simply, our minds go into a feedback loop: A tsunami is so far beyond our normal day-to-day experiences that we freeze—the "deer in the headlights" syndrome if you wish. You stand there, not moving until it is too late—and you die.

If a tsunami comes from a long way off and there are hours of warning from a modern-day warning system that works, then you should survive. However, what happens if you are a few minutes away from the source? Only if you have a deep understanding of the sea, are aware of an approaching tsunami, or know what signs to look out for and what action to take do you really have any chance. Landslide-generated tsunamis can be particularly grim because they might just happen—no warning except for the sea suddenly rising up or going out a longer distance than usual before it comes back in to shore. Miss these signs and you have no time to escape. Such events happen frequently, and yet we generally hear nothing about them. They often hit shores with few people or are mistaken for some other type of rogue wave.

The Taan Fjord tsunami of October 2015 is one of at least eight similar events that have occurred in Glacier Bay National Park, Alaska, since 2012. The wave reached more than 600 ft (192 m) above sea level. No one was killed; no one was around. As the old saying goes, "If a tree falls in the forest, does anybody care?" To paraphrase this, "If a catastrophic tsunami occurs somewhere with few or no people around, does anybody care?" They should—with climate change comes some serious problems. For example, glaciers retreat, leaving beautifully picturesque fjords with deep water and steep sides just right for cruise ships to enter. But beware, the valley sides are no longer propped up by ice and so they can easily fall into that deep water and generate some very nasty tsunamis. Read on because next we present the story of the largest historical event ever recorded.

This was truly *big*.

5

The Monster of Lituya Bay

> And so the legend of Lituya tells of a monster of the deep who dwells in the ocean caverns near the entrance. He is known as Kah Lituya, "the man of Lituya." He resents any approach to his domain, and all of those whom he destroys become his slaves, and take the form of bears, and from their watch towers on the lofty mountains of the Mt. Fairweather range they herald the approach of canoes, and with their master they grasp the surface water and shake it as if it were a sheet, causing the tidal waves to rise and engulf the unwary.
>
> **—Emmons (1911, p. 295)**

The Discovery

In a remote area of southeast Alaska, bordering British Columbia, lies a narrow, mysterious, and deadly arm of the sea called Lituya Bay. The Tlingit Indian legend tells of a monster that dwells in the bay and destroys all who enter his domain by grasping the surface of the water and shaking it as if it were a sheet. No Tlingit live in Lituya today. They left at an unknown time and for an unknown reason. But it must certainly have had something to do with the monster and the bay's reputation for evil.

Lituya Bay was first discovered by Europeans on July 2, 1786, when the French explorer, Jean Francois de Galoup, Compte de la Perouse, sailed his vessels *Astrolabe* and *Boussole* through a narrow channel and into the bay. La Perouse named it Port des Francais, and 19th-century whalers called it Frenchman's Bay.

This French expedition largely took place because the French had been one-upped by Captain James Cook who, on his third (and final) voyage into the Pacific, had put so many places on the map, literally, that France wanted some of the action. In a single visit in 1778, Cook charted the majority of the North American northwest coastline, mapped the full extent of Alaska, and identified what is now known as Cook Inlet, which today is home to Anchorage—we explore this approximate area in the Gulf of Alaska in Chapter 8.

So, the French decided to mount an exploration to rival Cook's efforts. The expedition left France in August 1785 and spent the summer of 1786 off the coast of Alaska searching for a northwest passage—a common theme of expeditions in those days.

Lituya Bay is plagued by a narrow and difficult entrance, nearly closed by a glacial moraine forming a spit called La Chaussee, so named because it reminded the Frenchmen of a causeway. As La Perouse's two small ships approached the entrance, they had their first indication of the welcome Lituya held for visitors. When his vessels attempted to sail through the narrow pass at the mouth of the bay, they were whirled completely out of control. La Perouse would write about the experience, "In my 30 years of navigation, I never saw two ships so near destruction." Once safely at anchor inside the bay, La Perouse sent three boats to carry out a survey of the entrance. The boats were seized by breaking waves in a tidal bore and swamped. The great navigator lost 21 of his men to Lituya. He erected a monument, a cenotaph, to the memory of the dead sailors on a small, round island in the bay, to this day known as Cenotaph Island. La Perouse left Lituya as quickly as possible. For him, the bay had an aura of evil.

The Bay

The bay is T-shaped, with the stem of the "T" cutting through the coastal lowlands and foothills flanking the Fairweather Range of the St. Elias Mountains (Figure 5.1). The stem is approximately 9 miles long and ranges from 0.75 to 2 miles (1–3 km) wide, except at the entrance, which has a width of less than 1,000 ft (300 m) at low tide. Cenotaph Island divides the central part of the bay into two channels. The crosspiece at the top of the "T" is approximately 3 miles (5 km) long and terminates in the nearly perpendicular faces of the Lituya Glacier to the north (Gilbert Inlet) and Crillon Glacier to the south (Crillon Inlet). The Fairweather Fault lies in this crosspiece, covered by glaciers and up to 500 ft (150 m) of water in the inlets. The area where the stem joins the crossbar of the "T" is flanked by two sharply rising promontories.

The Lituya and North Crillon Glaciers, each approximately 12 miles (19 km) long and 1 mile (1.6 km) wide, originate in ice fields at 4,000 ft (1,200 m) and higher near the crest of the Fairweather Range. Around the head of the bay, the rock walls are steep and fjord-like, rising to altitudes of between 2,200 ft (670 m) and 3,400 ft (1000 m) in the foothills to the north and south and to more than 6,000 ft (1,800 m) in the Fairweather Range less than 2 miles from

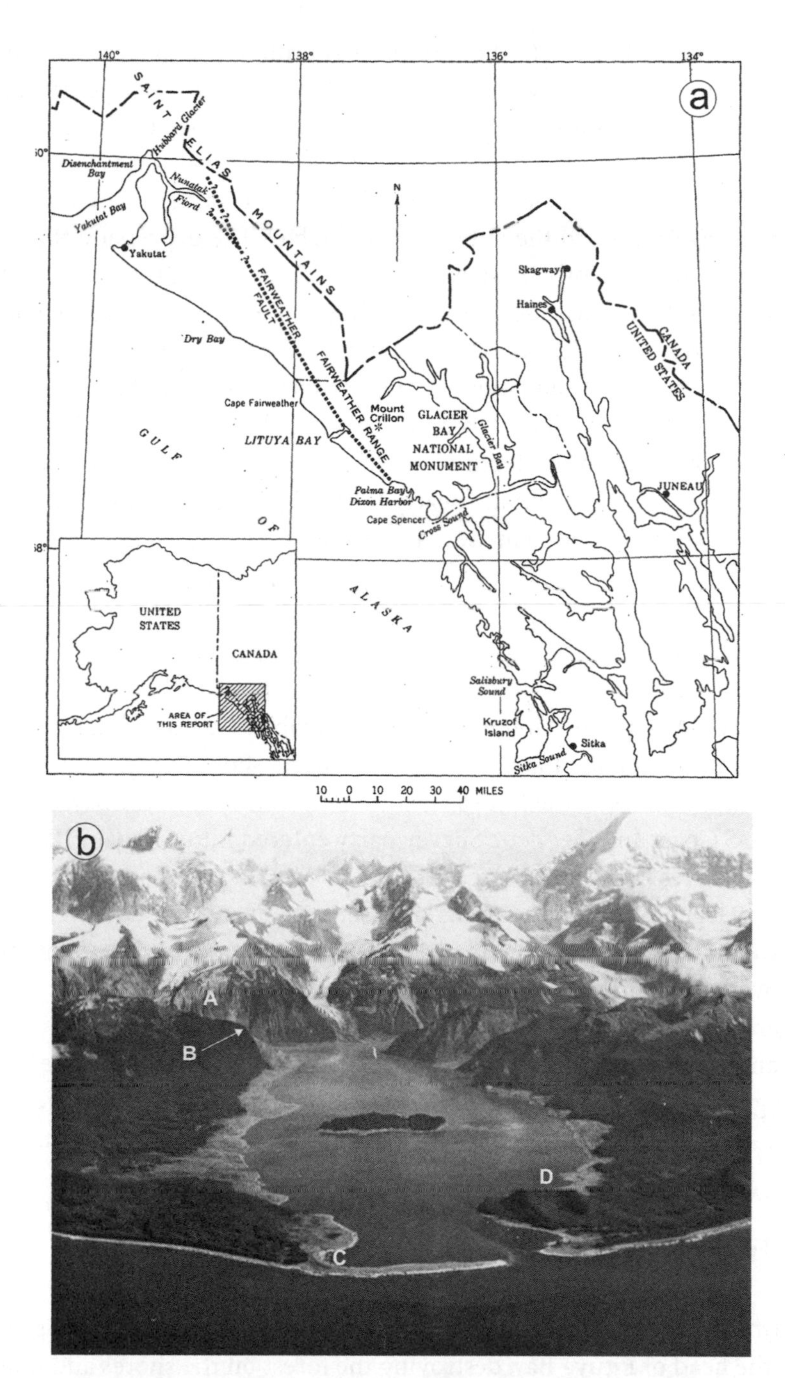

Figure 5.1 Lituya Bay and adjoining glaciers: (a) Regional map and (b) aerial photograph looking east–northeast. A, Rockslide source for tsunami; B, maximum run-up height of wave marked by destroyed forest; C, a fishing boat anchored in the cove here was carried over the spit in the foreground; and D, a boat anchored here rode out the wave (another boat underway near the entrance was sunk).

Source: Don J. Miller, US Geological Survey.

the shores of Crillon Inlet. The forests surrounding the bay contain spruce, hemlock, and cedar, with some trees probably more than 1,000 years old.

The submarine contours of the inlets show a pronounced U-shaped trench with steep walls and a broad, flat floor sloping gently downward from the head of the bay to a maximum depth of 720 ft (220 m) just south of Cenotaph Island and rising again toward the outer part of the bay. The depression that forms the bay had only recently been exposed as a larger Lituya Bay glacier retreat. The present Lituya, Cascade (at the crossing of the "T"), and Crillon Glaciers are the only remnants. The maximum extent of the Lituya Bay glacier is clearly recorded by the arcuate end moraine that forms La Chaussee Spit. This is what constricts the entrance of the bay and also makes for a shallow and dangerous channel with a minimum depth of only 33 ft (10 m) and a tidal current that can reach a velocity of 12 knots (nearly 14 miles per hour), thus making it difficult for a sailing ship to maneuver at low tide.

The Tlingit

The Native American Tlingit tribe had inhabited Lituya long before La Perouse's ill-fated encounter. Old Russian charts show a large Tlingit village of some 200 natives of the Lituya clan near the entrance to the bay, yet by the time a US Coast and Geodetic Survey party entered Lituya Bay in 1874, there were no natives living anywhere in the bay.

Tlingit legends tell stories of this large settlement and one day after the men had been out hunting sea otters, they returned to find the village completely destroyed, gone, literally washed away. There was only a single survivor—a women who had been picking berries high up on a hill.

In another story, two hunters in the mountains looked down on the bay and saw a giant flood pour down between the mountains and destroy the village. Yet another account speaks of seven or eight canoes capsized and lost.

The Incidents

Five times during slightly more than a century, giant waves have rushed out from the head of Lituya Bay, destroying the forest on the shores and hills and leaving trimlines high among the trees marking the height of each tsunami's destruction. The dates for these most recent events and the elevations of their forest trimlines are listed here:

1853/1854	395 ft (120 m)
1874	80 ft (24 m)
1899	200 ft (60 m)
October 27, 1936	490 ft (150 m)
July 9, 1958	1,720 ft (524 m)

Other than the evidence of the forest trimlines, little is known of the events of 1874 and 1899. Tree ring evidence indicates that the wave of 1853/1854 probably occurred between mid-August 1853 and May 1854. The wave almost certainly inundated the sites of villages described by La Perouse and may have destroyed the villages, as related in native oral histories. The events of 1936 and 1958, however, were observed firsthand by witnesses, some of whom lived to tell their tales.

The Monster Strikes in 1936

On October 27, 1936, four people observed an enormous wave at Lituya Bay. A fisherman named John Huscroft had been living on Cenotaph Island in the middle of Lituya Bay almost continuously since 1917. Another fisherman, Bernard Allen, had recently begun working with Huscroft and was sharing the rustic cabin, built 50 ft (15 m) above the bay, on the western side of the island.

Two other men were in Lituya that day on the fishing boat *Mine*, riding at anchor off the northern shore of the bay, approximately 1.5 miles (2.5 km) west of the cabin. According to the fisherman on the *Mine*, at approximately 6:20 a.m., 2 hours before sunrise, a dull, continuous rumbling was heard aboard the vessel. It appeared to come from the mountains located beyond the top of the bay. Because of the darkness, they could not see what was happening. No tremors were felt. The tide was rising and was at about mean tide at the time. The rumbling lasted until approximately 6:50 a.m., when the first large wave appeared in the narrowest part near the head of the bay. It was like a steep watery wall, stretching from shore to shore and having a possible height of 100 ft (30 m). Seeing this wave, the fishermen weighed anchor and headed for Cenotaph Island. When the wave reached them 10 minutes later, they were 1,300 ft (395 m) northeast of the island, in a depth of approximately 70 ft (21 m) of water. The arrival of the first wave was not preceded by a drop in water level but, rather, lifted the ship approximately 50 ft (15 m) higher.

Immediately after the passage of this wave, the water surface dropped below its normal level.

Huscroft's own seining boat, riding at anchor off Cenotaph Island in water 48 ft (14 m) deep, touched bottom. The first wave was followed by a second and third at approximately 2-minute intervals, and each successive wave was larger than the preceding one. After each of these waves, the surface of the water again fell below the normal level. A few small waves were observed during the half hour after the passage of the third wave, with all directed toward the exit from the bay. Following this, floating tree trunks and pieces of ice appeared in the water around the *Mine*.

Meanwhile on the island, the fishermen in the cabin were awakened at approximately 7:00 a.m. by the noise, "like the droning of a hundred airplanes flying at low altitude," and found that a wave was approaching their cabin. Climbing to a high, safe place, they saw three waves pass the island at a speed of approximately 25 miles per hour (40 km per hour) and growing in height. The maximum height of the waves was estimated at between 150 and 250 ft (45–76 m). The cabin was flooded, 50 barrels of salted salmon were washed away, and two wood frame outbuildings were destroyed. It was later determined that the water had risen to a height of 490 ft (150 m) along the northeast shore of Crillon Bay. The destruction of the forest extended to a maximum distance inland of approximately 2,000 ft (600 m) from the shoreline.

There was no earthquake that day, but the waves had occurred during a period of unusually heavy rainfall, so a rockslide or an avalanche is a possible cause. Also, sudden surging of the glacier or a submarine landslide are possible candidates to have produced the giant waves. No one knows. However, we know far more about the next event.

The Monster's Greatest Attack: The Events of July 9, 1958

Don Miller, a geologist working for the US Geological Survey, studied Lituya Bay in the early 1950s. Miller presented his data on the forest trimlines at the Annual Meeting of the Geological Society of America in Seattle in 1954 and hypothesized that giant waves were responsible. Oceanographers and hydrographic surveyors doubted his information and interpretation. They thought they knew the sea, but they did not know the "Monster of Lituya."

Four years later, though, his hypothesis came back to haunt them.

On the evening of July 9, 1958, three fishing boats, each with two people aboard, were anchored in the outer part of Lituya Bay just inside the spit. At

10:16 p.m. local time, about sunset at that time of year, a major earthquake occurred. Large trees in the area had their tops snapped off or were felled entirely. The earth cracked and heaved, and sand shot up out of holes in the ground. Many fissures appeared; one at the Akwe River, northwest of Lituya, was so large that a truck fell into it and was lost.

Elsewhere in Alaska, three people in a cabin on the East River were thrown to the floor by the force of the earthquake. They escaped from the building just before it was completely demolished and had to make their way to safety through a virtual forest of mud, sand, and spouting water as fissures were being formed.

Mariners approximately 12 miles (19 km) off Icy Point (~25 miles southeast of Lituya Bay) likened their experience to "riding on top of a big explosion." From their vantage point, they were able to hear loud roars as the mountainside was breaking away. Meanwhile, fishermen in Disenchantment Bay, near the head of Yakutat Bay, heard the "deafening roar of the ice breaking and the scarp-like glacier fronts cascading into the already churning waters."

Almost exactly 100 miles (160 km) from Yakutat lay Lituya Bay, the head of which was only 13 miles (20 km) from the epicenter of the earthquake. Here, the weather was clear with high scattered clouds, and the head of the bay was clearly visible from the deck of boats in the outer part of the bay.

Account of the *Edrie*—from Howard G. Ulrich

Howard G. Ulrich and his 7-year-old son, Junior, entered the bay on the *Edrie* at approximately 8:00 p.m. and anchored in 30 ft (9 m) of water in a small cove along the south shore. Ulrich was awakened by violent rocking of the boat and went out on deck to watch the effects of the earthquake. Approximately 2½ minutes after he first felt the earthquake, he heard a deafening crash coming from the head of the bay. "It was not a wave at first. It was like an explosion. . . . The wave came out of the lower part, and looked the smallest part of the whole thing."

Midway between the head of the bay and Cenotaph Island, the wave appeared to be a straight wall of water possibly 100 ft (30 m) high, extending from shore to shore. The wave was breaking as it came around the north side of the island, but on the south side it had a smooth, even crest. As it approached the *Edrie*, it appeared very steep and 50–75 ft (15–23 m) high. Ulrich continued to watch the progress of the wave for approximately 3 minutes until it reached his boat. He was unable to get the anchor loose, so he let out all 240 ft

(73 m) of chain and started the engine. Then the anchor chain snapped, and the boat began to rise with the wave.

Ulrich grabbed his radiotelephone and yelled, "Mayday! Mayday! This is the *Edrie* in Lituya Bay. All hell has broken loose in here. I think we've had it. Good-bye."

The boat was carried toward and probably over the south shore, and then, in the backwash, toward the center of the bay. After the wave passed, the water surface was very turbulent with much sloshing back from shore to shore and with steep, sharp waves up to 20 ft (6 m) high. Amazingly, after the first giant wave passed, Ulrich managed to get the boat under control and drove it out of the entrance of Lituya Bay at 11:00 p.m.

Account of the *Badger*—from William A. Swanson

Bill and Vivian Swanson of Auburn, Washington, had entered Lituya Bay at approximately 9:00 p.m. on their boat, the *Badger*. They first went as far as Cenotaph Island and then returned to Anchorage Cove on the north shore near the entrance to anchor in approximately 25 ft (7 m) of water near another fishing trawler, the *Sunmore*.

Bill was awakened by violent vibration of the boat. He looked toward the head of the bay, past the north end of Cenotaph Island, and saw what he thought to be the Lituya Glacier, which had "risen in the air and moved forward so it was in sight. . . . It was jumping and shaking. . . . Big cakes of ice were falling off the face of it and down into the water." After a while, "the glacier dropped back out of sight and there was a big wall of water going over the point" (the spur southwest of Gilbert Inlet).

As the wave passed Cenotaph Island, it seemed to be approximately 50 ft (15 m) high near the center of the bay and sloping even higher toward the sides. It passed the island approximately 2½ minutes after it was first sighted and reached the *Badger* approximately 1½ minutes later. The *Badger*, still at anchor, was lifted up by the wave and carried across La Chaussee Spit, riding stern first just below the crest of the wave, like a surfboard, only backwards. The Swansons looked down on the trees growing on the spit and believed that they were approximately 80 ft (24 m) above the treetops! The wave crest broke just outside the spit, and the boat went down almost vertically with her bow in the air. She hit bottom and foundered some distance from shore.

Somehow, Bill and Vivian managed to launch their 8-ft (2.5-m) dingy and climb into it. The dingy had been nearly swamped and their oars washed away, but Bill ripped loose the seat and with that as an oar began paddling. They were rescued 1½ hours later, waist deep in icy water, cold and in shock, by Julian Graham aboard his fishing boat, *Lumen*. Graham had overheard radio traffic from Howard Ulrich's "Mayday" and rushed toward the bay to render help.

Account of the *Sunmore*

Orville and Mickey Wagner from Idaho Inlet were also anchored in Lituya Bay on their 55-ft (16-m) trawler, the *Sunmore*. Though anchored near the *Badger*, when they saw the wave, they made a run for the entrance to the bay, trying to get outside before the wave could overtake them. They didn't make it. The wave lifted and tumbled them over the south side of the entrance, and the boat went down in the ocean just outside the bay. No trace of the vessel or the Wagners was ever found.

A Close Call

A party of eight Canadian mountain climbers had been camped in tents on the shore of Anchorage Cove in Lituya Bay, at the base of La Chausee Spit. They had just returned from the second-ever ascent of Mount Fairweather, British Columbia's highest peak and a boundary with Alaska. They left at approximately 8:00 p.m. on the evening of the quake in an amphibious plane. This was a day earlier than planned because their pilot was worried about the weather. They took off slightly more than 2 hours before the wave washed over their campsite.

Also in the area, geologist Virgil Mann of the University of North Carolina and a party of 16 men were camped on the shore of Lake Crillon, 8 miles southeast of Lituya. They had been planning to move to Huscroft's abandoned cabin on Cenotaph Island the very next day.

After the Wave

On the morning of July 10, Gilbert and Crillon Inlets and the upper part of the main trunk of Lituya Bay were covered by an almost solid sheet of floating ice blocks, some as large as 50 × 100 ft (15 × 30 m). Logs were scattered throughout the rest of the bay and over a fan-shaped area of the sea as far as 5 miles (8 km) from the entrance of the bay.

During the earthquake, approximately 1,300 ft (400 m) of ice had been sheared off the glacier. The most striking change at the head of Lituya Bay, apart from the new trimlines, was the fresh scar on the northeast wall of Gilbert Inlet, marking the recent position of a large mass of rock that had plunged down the steep slope into the water, almost certainly triggered by the earthquake of July 9.

The trimline formed by the 1958 wave extended to a maximum height of 1,720 ft (524 m) above sea level on the spur southwest of Gilbert Inlet—that's approximately 100 m higher than the Empire State Building. Its maximum horizontal distance was some 3,600 ft (1,100 m) from the high-tide shoreline. The total area inundated by the wave was at least 5 square miles (8 square kilometers).

One of the most impressive aspects of the 1958 wave is the thoroughness of its destruction of the forest. In most places, the trees were washed out and carried away, leaving bare ground. At Harbor Point, the easternmost point of the entrance to Lituya Bay, a living spruce tree was broken off cleanly approximately 3 ft (1 m) above the root system, where it measured 4 ft (1.2 m) in diameter. Many of the trees felled by the wave were reduced to bare stems, with the limbs, roots, and even bark removed by the turbulent water. A minimum depth of at least 1 ft (30 cm) of soil was removed over the entire area, amounting to more than 4 million cubic yards (3 million cubic meters) of earth.

No trace could be found of the Huscroft's cabin on Cenotaph Island or of a lighthouse mounted on concrete piers at Harbor Point.

Nature of the Wave

It was movement along the Fairweather Fault that caused the earthquake. At one point on the fault, the southwest side moved laterally (northwest) by at least 21.5 ft (6.5 m) and vertically by 3.5 ft (1 m). The shaking was extremely violent and probably lasted for at least 1 minute. The earthquake resulted in a large mass of rock [2400 × 3000 × 300 ft (730 × 910 × 91 m), some 40 million cubic yards (30 million cubic meters), weighing approximately 90 million tons (81.5 million metric tons)] sliding from the northeast wall of Gilbert Inlet. The movement of the rock was accompanied by clouds of rock dust and avalanching snow and ice, which may account for the "jumping glacier" seen by the Swansons. The impact of the large rock mass into the water produced a deafening crash and caused a huge sheet of water to surge up over the spur on the opposite side of Gilbert Inlet to a height of 1,720 ft (524 m). The impact

Figure 5.2 Trunk of tree [~4 ft (1.2 m) in diameter] sheared off by one of the tsunami waves.
Source: Don J. Miller, US Geological Survey.

also set in motion a tsunami wave with a steep front traveling south at speeds between 97 and 130 miles per hour (156–209 kilometers per hour) (Figure 5.2). Even after the wave passed, the bay was still in violent turbulent motion with waves surging from shore to shore for at least 25 minutes.

Epilogue

Don Miller returned to Lituya Bay following the 1958 event in order to survey the impact of the giant waves. By documenting the tsunami, he proved his theory correct. Miller's work in Lituya Bay helped greatly increase our understanding of tsunamis caused by landslides. Six years later, the magnitude 9.2 Great Alaska earthquake would trigger landslide tsunamis across southern Alaska, accounting for many of the deaths from that earthquake (see Chapter 8). Unfortunately, Miller was not around to study that one; he died on August 6, 1961, while trying to rescue some of his research team who fell through the ice on the Kiagna River in Alaska. Fittingly, a research vessel named after Miller served as the base of operations for some of the work studying changes to the Alaska coastline following the 1964 earthquake. This work would greatly advance our understanding of subduction zone earthquakes, the generators of ocean-wide tsunamis.

No one has lived permanently either in or near Lituya Bay since John Huscroft died in 1940. In recent years, the bay has come into increased use as an overnight anchorage and refuge in bad weather for the fishing boats of the fishermen who ply the adjoining waters of the Gulf of Alaska during the summer and early fall. But the monster is not dead; it is merely resting, regaining strength, and waiting.

Currently, adventure tourism is tempting fate. Backcountry hiking trips now pass along La Chaussee Spit, and weary hikers rest a while as they wait to be ferried across the entrance to the bay to continue their awe-inspiring journey. Let's hope they make it.

The entrance is only approximately 550 yards (500 m) wide at this point—the bay is almost a lake, but not quite. But this raises an interesting question: Can these things happen in lakes too? And if they can, then what about dams and rivers and . . . anything freshwater? We discuss freshwater tsunamis next.

6

The Perils of Freshwater Tsunamis

> The great mountain of Tauretunum, in the territory of the Valais, fell so suddenly that it covered a castle in its neighborhood, and some villages with their inhabitants; it so agitated the lake for 60 miles in length and 20 in breadth that it overflowed both its banks; it destroyed very ancient villages, with men and cattle; it entombed several holy places, with the religious belonging to them. It swept away with fury the Bridge of Geneva, the mills and the men; and, flowing into the city of Geneva, caused the loss of several lives.
>
> **—Marius of Avenches (cited in Bonney, 1868, p. 131)**

Who Cares?

There are many examples of freshwater tsunamis, but they tend to get forgotten in the litany of horrific events that have come from the sea. Although there are numerous examples of freshwater tsunamis, few scientific studies of them have been carried out. Why is that?

It is often difficult for people to get their heads around freshwater tsunamis, but consider the Lituya Bay event. Yes, that was a landslide into a bay, but that bay was almost completely enclosed with a narrow entrance to the sea only 0.31 miles (500 m) across. In other words, it is almost a lake.

As a minor aside (because this chapter is about freshwater tsunamis), although scientists spend much time impressing people, or scaring them, with details about how large the 1958 Lituya Bay event and earlier ones were, these have only ever been recognized by examining old trimlines of trees and doing tree ring dating. A trimline in this instance refers to the height to which the forest was either destroyed or badly affected by the wave (hence you can take a core of a tree and look for what is termed reaction wood where the tree suffered a serious upset). Scientists have used a variety of models to unravel the mystery of how these tsunamis, or at least the 1958 tsunami in Lituya Bay, were generated, but no one has ever gone there to study the geological evidence. Many may say that this is not needed. Perhaps, but this can be countered by asking questions such as "OK, so how big and how often do these things occur?" Or another perhaps more esoteric question might be, "What can we learn from each event about the power of the water, its ability to erode the earth and trees, and where did it deposit all of the stuff it eroded?" This type of information is crucial to understanding what occurred not only at Lituya Bay but also at similar sites throughout the world where many more people may reside. This becomes increasingly relevant in colder coastal areas, where ports and communities rely almost entirely on sea trade for their existence (e.g., northern Canada and Norway). In these areas, climate change is causing the rapid retreat of glaciers, leaving behind deep water-filled valleys with massive steep valley sides no longer supported by the glacier that used to fill them up. The valley sides sit there, exposed to the weather and the inevitable effects of gravity that will cause them to collapse into the water below.

In this context, the tag line of the Norwegian film, *Bølgen* (*The Wave*], is very appropriate: "Det har skjedd før. Det vil skje igjen"—"It has happened before. It will happen again." Or consider the ongoing monitoring of the approximately 3,000-ft (900-m)-high Åknesfjället Mountain that looms over a Norwegian fjord: It has a crack 2,300 ft long and up to 100 ft wide (700 × 30 m) that continues to grow at approximately 6 inches (15 cm) each year. At some point, the mountain will fall into the fjord, triggering a tsunami that is expected to be up to 260 ft (80 m) high. Located in the fjord is the small village of Geiranger with a population of approximately 200, and the village's seasonal tourist trade draws 800,000—1 million visitors per year, mostly between May and September—gulp.

A classic example of such a scenario—fortunately not involving people—happened in Taan Fjord, Alaska, on October 17, 2015, near the terminus of Tyndall Glacier (Figure 6.1). Here, a huge landslide of more than 2 billion ft^3 (~60 million m^3) of rock fell into the fjord. The resulting tsunami reached elevations as high as 630 ft (193 m). The Taan Fjord event was studied by geologists, who discovered much about it. This therefore begs the question as to why no one has studied Lituya Bay, where we know there have been several historical events. These type of events may only be locally significant, confined to the individual bays in question, but that simply means that all the energy of the tsunami will be expended in a small space and not dissipated over a large area.

And now back to freshwater tsunamis. Why not have tsunamis in lakes, rivers, or even possibly man-made structures such as reservoirs behind dams? After all, as examples such as Lituya Bay and Taan Fjord demonstrate, really only a landslide that falls into the water is needed to generate a tsunami, or even a large underwater slide—there are many reasons why freshwater

Figure 6.1 This photograph was taken high up on a mountainside where the Taan Fjord tsunami reached on October, 17, 2015. The slope seen in the distance across the fjord collapsed, sending a wave raging up the mountainside to approximately 190-m elevation not far from where this photo was taken.
Source: Ground Truth Trekking.

tsunamis could and have happened. Furthermore, there are many millions of people exposed to these events—they live either around lakes (enclosed environments) or along the banks of rivers (semi-enclosed), and they don't even know about the possibilities of a tsunami literally in their backyard.

Think of a decent-sized lake surrounded by hills and there may well have been a tsunami there in the past, although unfortunately, scientists have not spent much time searching for them. There are two probable reasons for this. First, in general, scientists are focused on saltwater tsunamis, which are by far the most prevalent type (although that could be a circular argument because not much research on their freshwater counterparts has been undertaken). Second, there is little available funding for it. After all, massive events such as the 2004 Indian Ocean and 2011 Japan tsunamis are devastating, and both they and their predecessors demand urgent study. When studies of less devastating events are carried out, they invariably focus on the enigmatic, "sexy," events such as Taan Fjord that are the hot favourites.

A classic example of a freshwater tsunami is that which occurred in Lake Tahoe in the Sierra Nevada Mountains, straddling the border between California and Nevada. Here, a massive 3-cubic-mile (12.5 km^3) landslide, probably generated by an earthquake on a local fault, fell into the lake and caused a huge tsunami, perhaps 330 ft (100 m) high. Some of it rushed out of the lake outlet by Tahoe City and cascaded down the Truckee River. House-sized boulders found as far downstream as Verdi at the border with Nevada are testament to the force of this event. When did it happen? Approximately 50,000 years ago. It is important to not heave a sigh of relief and think, "Oh, that's fine; that was ages ago." There are two things to remember here. First, this is the only one we currently know about at Lake Tahoe; there could be more as yet undiscovered. Second, it will happen again, but the key is when, and we do not know the answer to that. Another lake affected by at least one tsunami is Lake Tarawera in New Zealand. Here, as opposed to a landslide, the tsunami was caused by a pyroclastic flow—a fast-moving current of hot gas and tephra moving rapidly [62–430 mph (100–700 km/h)] down the side of a volcano. The Kaharoa eruption of Mount Tarawera in approximately 1315 AD produced numerous pyroclastic flows that entered Lake Tarawera. Coarse pebble deposits on the far side of the lake indicate a maximum tsunami wave height of approximately 23 ft (7.0 m). This may pale in significance compared to the Lake Tahoe tsunami, but it was still catastrophic. Fortunately, it seems likely that there was no one around at the time, but it will happen again.

Destruction of the Fortress of the God of Thunder

The Iron Age fortress Taranais dunum (*dunum* is derived from the Celtic "dunon," or "fortress"; *Taranais* was named after Taranis, in Celtic mythology, the god of thunder) eventually succumbed to Roman invasion from the south, and by the sixth century its name had been Romanized to *Tauredunum*. This fortress, located in an area prone to thunderstorms, hence the name, was on a strategically sited hilly plateau above the fertile plain of the River Rhône at the head of Lake Geneva. This was a good stronghold and a place from which military forces could control much of the surrounding region. It was almost impregnable, or at least it was to most human efforts, but then in 563,

> A great wonder appeared in Gaul at Fort Tauredunum, located on the river Rhone on the side of a mountain. For more than sixty days an indescribable roar was heard, then eventually a mountain fractured and separated from the next, collapsing into the river, carrying people, churches, property and houses. The river course became blocked and the water flowed backwards, flooding the area upstream, inundating and destroying everything on the banks. (Gregoire de Tours, 563)

The fort was destroyed, buried by debris from the fractured mountain—Les Dents du Midi. The landslide blocked the course of the river, and evidence has been found that shows that the whole of Saint-Maurice [7.5 miles (12 km) upstream from Lake Geneva] was covered in water that had backed up behind this blockage. A skeleton found deeply buried here in the gravelly sediments deposited during this event may well have been a victim. Indeed, it was not just Saint-Maurice that was flooded. Many villages were flooded either by the river as it backed up behind the landslide debris or when the river finally burst through this temporary dam, flooding downstream toward Lake Geneva.

Until recently, it was thought that it was the tsunami from the dam burst that engulfed settlements along the banks of Lake Geneva. However, workers from the University of Geneva now think that it was not the landslide that caused the tsunami, nor was it the dam burst that followed, but rather that the impact of the Tauredunum landslide as it hit the ground destabilized sediments at the mouth of the Rhône, causing them to collapse and flow rapidly into the lake, triggering a large tsunami.

A giant turbidite fan (sand and mud deposited by a rapid flow of water) extends over 6 miles (10 km) into the lake. This fan is approximately 3 miles (5 km) wide and 16 ft (5 m) thick, giving it a volume of at least 8.8 billion ft^3 (250 million m^3). The rapid collapse of this material into the lake would have generated a massive tsunami. But how massive? The Swiss team ran some computer simulations to determine just how large it would have been. This turbidite would have generated a wave up to 52 ft (16 m) high, and just as the historical accounts of the time describe, it would have traveled the full length of the lake within 70 minutes. Lausanne, approximately 9 miles (15 km) down the lake, would have been struck by a 43-ft (13-m) wave 15 minutes after the tsunami was generated. Lausanne was fortunate, however, because the shoreline is steep here and so flooding was not too bad, unlike Geneva at the very far end of the lake. Here, a 26-ft (8-m) wave would have struck the city after 70 minutes, causing widespread destruction on its way at intervening towns and villages—just as described by Marius of Avenches (cited in Bonney, 1868):

> It so agitated the lake for 60 miles in length and 20 in breadth that it overflowed both its banks; it destroyed very ancient villages, with men and cattle; it entombed several holy places, with the religious belonging to them. It swept away with fury the Bridge of Geneva, the mills and the men; and, flowing into the city of Geneva, caused the loss of several lives. (p. 131)

This catastrophic event devastated the region, but that is not the end of this story. The Swiss team that produced these new results also found evidence for four older turbidites. This type of sediment collapse has happened repeatedly in the past, although we don't know when exactly except that it must have been since the glaciers retreated after the last glaciation here approximately 19,000 years ago. However, the message here is clear: It has happened before and will undoubtedly happen again. Unlike the 563 Tauredunum event, however, the impact of a similar-sized tsunami in the future would be far more devastating because now there are well over 1 million people living along the lake's shores. And just to add one more wrinkle to the story, almost the entire southern side of Lake Geneva, approximately 22 miles (35 km) of this side of the lake, is composed of steep-sided mountains all capable of falling into the lake, so there is not just one tsunami source, but many. Which brings us back to the point about geological studies of these events. The Swiss work shows us clearly that there have been several other earlier events and we know how bad the most recent one was, so why is so little work being done in similar locations elsewhere throughout the world?

The Dam of Dishonor

The Lake Geneva story demonstrates that spectacular mountain scenery should come with a few important warning stickers: It is not only avalanches and broken legs while skiing that we need to be aware of—the threat of a tsunami is an ever-present risk, albeit not one that many ever contemplate, particularly skiers. So, what can we as humans do to make matters worse? Because, let's be honest about it, this is something that we are very good at.

We can, for example, build the "tallest dam in the world" in Italy. This was an idea hatched in the 1920s, but it was not completed until the late 1950s. It is still one of the tallest dams in the world today (Figure 6.2).

Figure 6.2 The Vajont Dam as seen from the village of Longarone in 2005, showing the top 60–70 m of concrete. The wall of water that overtopped the dam by 250 m would have obscured virtually all of the blue sky in this photo.

Source: Wikimedia Commons (https://commons.wikimedia.org/wiki/File:La_diga_del_Vajont_vista_da_Longarone_18-8-2005.jpg).

It was believed that the Piave and Maè Rivers together with the Boite stream, 60 miles (100 km) north of Venice, could be controlled by this dam and therefore pump prime industrial growth in the region by providing for the growing demand for power generation. It was called the Vajont Dam and was built by the Adriatic Energy Corporation [Società Adriatica di Elettricità (SADE)]. SADE pretty much had a monopoly on the electricity supply and distribution in northeastern Italy, and so it managed to buy the land despite opposition from the local communities downstream. SADE assured everybody that the geology of the mountain, including a lot of old landslides, was believed to be sufficiently stable for the dam to be built.

Construction work started in 1957 and was completed in late 1959, but over this time cracks and instabilities in one of the roads on the mountainside led to new geological studies being carried out. Three experts, all working separately, told SADE that the entire mountainside was unstable and was most likely going to collapse if the dam was filled. Filling the dam would cause water to seep into the base of these unstable rocks, reducing the friction and causing the rocks to collapse. As if to drive home this point, in March 1959 a landslide at the nearby Pontesei Dam generated a 66-ft (20-m) high tsunami that killed one person. And so . . .?

And so nothing.

The scientists were ignored, construction was completed in late 1959, and in early 1960 the company got the go-ahead to start filling the dam.

At the time of completion, the dam stood more than 850 ft (260 m) above the valley floor and could hold more than 220 million yards^3 (168 million m^3) of water.

But wait, there's more, as they say in TV infomercials.

As if the omens were not bad enough, minor landslides occurred throughout the summer of 1960, but essentially all the Italian government chose to do was sue at least one journalist for reporting these problems. Later that year, when the dam level had risen to approximately 620 ft (190 m) of the planned 860 ft (262 m), a 28 million ft^3 (800,000 m^3) landslide fell into the waters. In response, the water level was lowered by approximately 160 ft (50 m), and some engineering works were carried out to keep the dam operational.

And filling recommenced.

By late spring 1962, when the water level had reached 705 ft (215 m), there were two magnitude 5 earthquakes, but no significant landslides, and the reservoir continued to fill. However, there must have been some growing concerns about the stability of the site because several months later landslide models produced by the company's engineers indicated that such events could generate a tsunami capable of overtopping the dam if the water level was 66 ft

(20 m) or less from the top. As a result, the water level was initially held at 82 ft (25 m) from the top of the dam, but the engineers figured that they could control the rate of landslides by varying the water level.

So, they starting filling it up again—more water, more power, more money.

A year later, the dam was nationalized and came under the control of the National Agency for Electricity (ENEL) as part of the Italian Ministry for Public Works. The dam was full, but landslides and earthquakes continued and then in September 1963 the entire mountainside slid 8.7 inches (0.22 m) downhill—a mere blip, but the whole mountain was noticeably unstable. On September 26, concerned about this development, ENEL decided to start slowly lowering the water level by approximately 65 ft (20 m) to approximately 790 ft (240 m), but natural processes had overtaken whatever human decision-making was going on. A few days later, the entire mountainside moved almost 3.3 ft (1 m). Early on October 9, 1963, engineers saw trees falling and rocks rolling down into the lake where a predicted and modeled landslide could take place; the lowering of the water to a "safe" level was not working.

And then it happened.

At 10:39 p.m. on the same day, an approximately 9 billion ft^3 (260 million m^3) landslide composed of trees, sediment, and rocks fell into the reservoir at a speed of up to 68 mph (110 km/h). It lasted just 45 seconds and completely filled the reservoir behind the dam, rapidly displacing 1.75 billion ft^3 (50 million m^3) of water. With nowhere else to go but down, the massive wave of water overtopped the dam as an 820-ft (250-m)-high tsunami. In the Piave valley below, it destroyed the villages of Longarone, Pirago, Rivalta, Villanova, and Faè, killing approximately 1,900–2,500 people. US Army helicopters carried out more than 300 flights and ferried 4,000 survivors to safety. Although the destruction downstream was catastrophic, the dam was largely undamaged and, amazingly, it still stands today.

Immediately after the disaster, the government (which at the time owned the dam), politicians, and public authorities insisted on attributing the tragedy to an unexpected and unavoidable natural event.

However, apart from journalistic attacks and the attempted cover-up from news sources aligned with the government, there had been proven flaws in the original geological assessments and disregard of warnings about the likelihood of a disaster by SADE, ENEL, and the government.

A trial was inevitable.

Interestingly, the judges who heard the preliminary trial decided to move the main event to L'Aquila (Abruzzo) approximately 600 km (350 miles (600 km) south. This rather conveniently prevented local public participation and resulted in lenient sentencing for a few of the SADE and ENEL engineers.

Hindsight is a wonderful thing, but it was recognized that there were a number of problems with the choice of site for the dam and reservoir. The canyon was steep-sided, the river had undercut its banks, and the limestone and clay-stone rocks that made up the walls of the canyon were interbedded with the slippery clay units all sloping down toward the dam and toward the axis of the canyon. There were many other problems too, all making this a place *not* to build a dam.

Most of the survivors were relocated to a newly built village 31 miles (50 km) southeast, but some insisted on returning to the Piave valley. A pumping station was installed in the dam to keep water at a constant level, and the Piave River was essentially channeled through a tunnel to let it flow down into the valley. The reservoir is dry, filled by the landslide, and because of public interest the dam was partially opened in 2002. The tragedy became regarded as part of the price for economic growth in the 1950s and 1960s, but it is now part of a push to stimulate tourism in the area. As silver linings go, disaster tourism is not a great outcome.

It was inevitable that some type of movie would be made about this, and true to form, in 2001 a joint Italian and French docudrama about the disaster was released. It was titled *Vajont: The Dam of Dishonor*.

A Minor Aside: Should We Predict Earthquakes (and, Equally, Tsunamis)?

There is a certain reluctance on the part of seismologists to delve into this field. It is generally agreed among scientists that earthquake prediction is a "no win" situation. If you do predict an earthquake and nothing happens, you get blamed for the cost of preparation, evacuation, lost business income, and the needless disruption of people's lives. If you don't predict an earthquake and one does occur, you could be blamed for not predicting it. Are the earthquake scientists being oversensitive?

In 2009, a 6.3-magnitude earthquake struck the town of L'Aquila (where the Vajont Dam trial was held), killing 309 people. In 2012, six Italian seismologist were tried in an Italian court for not providing the public with acceptable earthquake risk information—in other words, for not predicting the earthquake. The scientists were found guilty and sentenced to 6 years in prison, in addition to having to pay millions in fines and damages. To quote a frequently asked question on the US Geological Survey's (USGS) website:

> Question: Can you predict earthquakes?
> Answer: No. Neither the USGS nor any other scientists have ever predicted a major earthquake. We do not know how, and we do not expect to know how any time in the foreseeable future. (https://www.usgs.gov/faqs/can-you-predict-earthquakes)

It appears that the father of modern seismology, Charles Richter, was correct when he supposedly said that only fools and charlatans predict earthquakes.

Rippling Waters

There is something almost hypnotic about rivers. For some people, it is possibly the best form of relaxation they can find, to simply sit on a river bank and watch the water with all of its eddies and ripples. Perhaps this is why the Māori of New Zealand gave the name Waikari (Rippling Waters) to the river that flowed into Hawke's Bay on the east coast of the North Island. Or perhaps this is a hint of a more sinister provenance, something slightly unstable about what should be the smooth-flowing waters of a river reaching the end of its journey to the sea.

In 1931, the region of Hawke's Bay wasn't exactly a sleepy backwater. In fact, the famous explorer Captain James Cook was one of the first Europeans to see the future site of Napier when he sailed down the east coast in October 1769. He observed, "On each side of this bluff head is a low, narrow sand or stone beach, between these beaches and the mainland is a pretty large lake of salt water I suppose" (https://www.napier.govt.nz/napier/about/history/). It looked like a pretty good place for future European arrivals to settle. However, the value of this region had been recognized long before by numerous Māori tribes, with Ngāti Kahungunu eventually becoming the most important group in the region. And then in 1851, the Crown purchased the Ahuriri block as it was known (including the site of Napier), and European colonization began.

Founded in 1855 by the government, Napier (formerly known as Ahuriri) was established on a small semi-island between the sea and an inner harbor. This was a perfect site for a port, and as a result, Napier became the leading town of the region, although lack of space for urban growth was a problem.

Approximately 80 years later, the space problem was solved.

On February 3, 1931, New Zealand's deadliest civil disaster in history struck. A magnitude 7.8 earthquake that struck at 10.47 a.m. killed 256 people (161 in Napier). Much of the city collapsed; fire broke out in the business district, and once the reservoir emptied, firefighters were powerless to put out the flames. The tragedy continued to unfold for many days as emergency supplies and help rushed to the region. Ultimately, the city returned to life and was resurrected literally from the ashes. In this case, there were some significant positive outcomes. The earthquake caused the surrounding land to lift up by almost 10 ft (3 m). This drained Cook's "large lake of saltwater" otherwise known as Ahuriri Lagoon, thus making a sizeable chunk of flat land right next to the city available for farms, industry, housing, and, eventually, Napier Airport. In addition, from an architectural perspective, the timing of the event meant that much of central Napier was rebuilt in an art deco style. As a result, the art deco style city sees an increasing number of tourists visit each year to marvel at its many buildings.

But what has this got to do with freshwater tsunamis?

The earthquake probably generated a small tsunami around Napier, although no conclusive geological evidence has been found. That would have been a saltwater one anyway. However, the earthquake also generated many landslides that blocked roads, destroyed infrastructure, and caused at least one tsunami, possibly more.

It seems strange to be so vague about an event that occurred less than 100 years ago when newspapers, photographs, and personal accounts of the time give us so much detail, but outside the larger towns and settlements of New Zealand, there were (and still are) many isolated farms dotted around the countryside. There was little or no media coverage in these sleepy backwaters, and so the only evidence is the records left behind by landowners or in the geology. It was at just such a location, the Waikari Station approximately 25 miles (40 km) north–northeast of Napier, that a record exists—although it must be said that there are two newspaper reports, almost unprecedented coverage for such a rural location. These stories, which were published well over 1 month after the event occurred, were based largely upon aerial reconnaissance flights and state near the mouth of the Waikari River a hillside burst open and collapsed into the river, causing it to sweep over Mr James Tait's home then situated high on the flats above; on the other side of the river Mr John Tait's woolshed and thousands of pounds worth of machinery, were swept away (Woodhouse, 1940, p. 54) and "the grass of the terrace was strewn with fish" (Anon, 1931a, p. 8).

There must have been a lot of fish if they could be seen so well from the air. Fortunately, we have the memoirs of the landowner at the time, Mr. Tait, and these are more informative and, not surprisingly, focused on the damage to his property.

Tait, who spent much of his life on Waikari Station and was there on the day the tsunami occurred, reported a wave approximately 50 ft (15 m) high: "It also lifted the waters of the river onto the top terrace, surrounding the homestead and washing some of the outbuildings a chain [66 ft or 20 m] or so away" (Tait, 1977; Waikari Station, 1840 to 1940, p. 79). Both the homestead [260 ft (80 m) from the river] and the sheep shed [100 ft (30 m) away] were destroyed, and "The river was completely blocked by this huge earth outfall" (Tait, 1977; Waikari Station, 1840 to 1940, p. 79). The debris of these buildings was strewn along the beach, 100 acres of land were swept out to sea, and sheep and cattle were engulfed by the waters.

Recent geological research at the site shows that a landslide of approximately 60 million ft^3 (1.7 million m^3) fell into the river, displacing the water and sending a large tsunami up onto the opposite bank, inundating Waikari Station.

Little did Mr. Tait know, but as wave height goes, this was and still is New Zealand's highest historical tsunami. And what evidence is there for it today? Intriguingly, given the importance of this event in New Zealand's tsunami history, it is surprising to discover that no research was carried out on it until 2015. We will never know what evidence has been lost in the intervening years, but what we do know is that all that could be found of the remains of the buildings were two roofing nails and a piece of broken crockery. Fortunately, some of the deposit remains, visible as a layer of river gravel covering the ground leading up to the former sites of the homestead and sheep shed.

This may seem like a paltry event when compared with some of the other events that we have discussed, but there is an important message here. There have been very few studies of freshwater tsunamis, but these tsunamis happen much more frequently than people imagine. The largest earthquake to hit the Hawke's Bay area before 1931 was a magnitude 7.5 in February 1863. Residents living in Hawke's Bay at the time reported numerous landslides, collapsed chimneys, and damaged houses. There were no reports from the remote Waikari Station, but the geology remains and there is evidence of an earlier tsunami probably caused by this 1863 earthquake, only 8 years after Europeans settled the area. Researchers had no time to look deeper, but it is pretty much certain that there are many others beneath those two, testimony to ongoing geological processes.

And that is the point: Tsunamis, whether they occur in seawater or fresh water, have happened much more often than people think. As we have seen from Papua New Guinea, humans have had a long-standing interaction with these catastrophic events and yet we forget. We always do.

It seems strange to be talking like this when we have both spent almost our entire careers studying some aspect of tsunamis. Have we failed in our mission to educate people about the things? Has all of our work been wasted?

We think not, but it is an uphill struggle. Like all hazards, we have to keep re-educating people. This is not surprising because we all have better things to do, other things to occupy our time, and in this day and age what many people consider to be of interest and newsworthy is what social media and the internet tell them is of interest and newsworthy. Sigh.

Let's try another approach. Time to appeal to sailors, who should have a healthy respect for tsunamis, but they too have experienced disaster.

7
Tsunamis and the US Navy

> The most famous stranded vessel is probably the USS *Wateree*, marooned on land by the AD1868 Arica tsunami in southern Peru (now northern Chile). It suffered from ongoing anthropogenic activities including use as a hospital and then a hotel, followed by use for target practice during the Guerra del Pacífico (War of the Pacific 1879–1883). It then suffered from scavenging, souvenir collection, and more recently, relocation of the last remaining piece, a boiler, closer to the coastal highway as a tourist attraction. Ironically, scavenging and souvenir collection was aided by the AD1877 Chilean tsunami that broke the stranded vessel into pieces during run-up and moving these remnants closer to the shore during the backwash.
>
> **—Goff (2012, p. 102)**

The military might of some of the world's largest countries is partly reflected in their naval prowess, perhaps none more so than the United States, but it is not always military.

Following the disastrous December 26, 2004, tsunami, more than 15,000 US sailors, soldiers, airmen, marines, and coast guardsmen, 24 US naval ships, and 1 Coast Guard vessel provided aid to the people of the regions stricken by the terrible event and distributed more than 2.7 million pounds (1.2 million kg) of relief supplies. However, this was not the first case of tsunami relief provided by the US Navy. Following the disastrous April 1, 1946, tsunami that struck the Hawaiian Islands, many of the survivors around the town of Hilo were sheltered at the US naval air station. A US flying boat aircraft made a valiant, though futile, attempt to save children washed out to sea by the tsunami (see Chapter 1), and a Navy tank landing ship rescued a women and a 15-year-old boy who had floated offshore for more than 24 hours. But the US Navy's involvement with tsunamis dates back much further in time to the days when it was first becoming a global sea power.

Tsunami in the Virgin Islands

During the US Civil War, Confederate blockade runners and privateers had turned the Caribbean into a lawless sea. Following the war, the United States wanted to put a stop to smuggling and privateering in the region and therefore wanted to have its own naval base in the area. By mid-November 1867, three US vessels had steamed into the Caribbean to evaluate sites for just such a naval base. The US Virgin Islands, at that time a Danish possession, seemed like an ideal location; moreover, rather conveniently, the King of Denmark was anxious to sell— for a mere $7.5 million. An envoy from the King had recently arrived on the island of St. Thomas (one of the US Virgin Islands) aboard the Royal Danish Mail Steamer *La Plata*, which was anchored in the harbor at the town of Charlotte Amalie. Also anchored in the harbor at Charlotte Amalie were the US Navy side-wheeled steamers USS *De Soto* and USS *Susquehanna*. Both vessels had served with distinction during the Civil War. The *Susquehanna* had fought in the famous battle with the Confederate ironclad *Virginia* (formerly the *Merrimack*) and had participated in the largest land–sea battle in history to that point, the assault on Fort Fisher, North Carolina, in February 1865. The *De Soto*, on the other hand, had served in the Gulf of Mexico during the Civil War, capturing 23 Confederate blockade runners.

Off the island of St. Croix some 30 miles (50 km) to the south was the USS *Monongahela*. Like the *De Soto* and *Susquehanna*, she was fresh from an illustrious naval career in the Civil War. The ship had been part of Admiral Farragut's squadron that ran past the Confederate batteries at Mobile Bay when the Admiral gained fame with his slogan, "Damn the torpedoes. Full steam ahead!" In this engagement, the *Monongahela* suffered 27 casualties, including her captain, but she later served as Farragut's own flagship and remained on duty in the Gulf of Mexico until the end of the war, stopping numerous blockade runners. Following the Civil War, she was assigned to the West Indies Squadron and hence she came to be anchored in the harbor of St. Croix on November 18, 1867.

On that day at 2:45 p.m., a violent earthquake, which would later be estimated to have had a magnitude of 7.5, occurred on the seafloor of the Anegada Trough between St. Croix and St. Thomas.

At St. Croix, onboard the *Monongahela*, Commodore Bissell stated,

> The first indication we had of the earthquake was a violent trembling of the ship . . . lasting some 30 seconds, and immediately after the water was observed

> receding rapidly from the beach; the current changed almost immediately, and bore the ship towards the beach.

When the ship was within a few yards of going up on shore, the current suddenly slackened. Commodore Bissell had the sails set and began to make for open water. "A light breeze from the land gave me momentary hope," said the Commodore. And then,

> the sea returned in the form of a wall of water 25 or 30 feet high [6–9 m], it carried her over the warehouses into the first street fronting the bay. The reflux of this wave carried her back towards the beach leaving her nearly perpendicular on a coral reef, where she has now keeled over to an angle of 15°. All this was the work of only some three minutes of time.

During the tsunami, an eyewitness on a nearby hill described the *Monongahela* as floating on the top of the great incoming wave on a level with the tops of the houses on Bay Street. The giant waves had been ruinous for the stores along the bayfront; they were inundated and their goods washed out and scattered. Fortunately, however, the authorities had quickly opened the jail, allowing the prisoners to flee for their lives.

Meanwhile at St. Thomas, the *De Soto* was anchored with the *Susquehanna* near the entrance to the harbor of Charlotte Amalie when, according to her captain, David Hall, they "experienced a heavy shock" at 2:50 p.m. Surprisingly, Captain Hall seems to have been prepared for a tsunami. It just so happened that his previous cruise had been with the European squadron, whose principal port was Lisbon, Portugal. He had become familiar with the Great Lisbon Earthquake and the tsunami of 1755. Captain Hall believed that the earthquake might be "followed by a tidal wave, so I kept a sharp lookout seaward." He goes on, "I did not have long to wait, for at 3:05 p.m. . . . I saw this immense wall of water." The "sea washed into the harbor about 20 feet high [6 m]" and the captain "called all hands to save ship." Standing just behind Captain Hall as the tsunami approached was Commodore Boggs, who was not quite so self-assured about the tsunami as the good captain. According to the commodore, the incoming wave had the "appearance of a great bore not less than 23 feet in height [7 m]." He continued, "The Island appeared to be sinking."

The captain had the anchors set, but the tsunami proved too much for the chains, which soon snapped and the ship began "drifting all over the harbor," the water "washing in and out with great force." The iron wharf just east of the *De Soto* was destroyed by the tsunami, and the ship was carried over the wharf and impaled on the broken pilings, punching three holes in the starboard side of her hull. Captain Hall wrote in his log, "ship leaking badly. . . . Released all prisoners." The captain finally managed to secure the ship at anchor and listed her to the port side to reduce the leaking.

Not far away, Admiral Palmer was on his flagship, the *Susquehanna*, in his cabin writing when the tremor struck, "accompanied by a sound resembling the grounding of a vessel upon a rough bottom." The admiral calmly resumed writing and "had been seated about ten minutes when the report was brought to me that the sea outside of the harbor had risen and was coming in a huge volume as if to engulf us all." The admiral went on deck in time to witness the tsunami wave approaching. "With a feeling of awe, we awaited its arrival; it came rushing on, tumbling over the rocks that formed the entrance, carrying everything before it." The *Susquehanna* managed to ride over "three waves in succession—the anchor chains holding on bravely." But from onboard they witnessed small craft in the harbor "lifted up and thrown into the streets. Boats were capsized and men swimming for their lives." As soon as the waters had calmed sufficiently to be safe, the men of the *Susquehanna* rendered aid to those still in danger, her boats picking up and saving several "drowning men."

Anchored near Water Isle just outside the main harbor, those onboard the *La Plata* could see the water "pile up into a wall as it approached the harbor." One of her passengers, Mr. William Maskell, reported that the first wave was a 40-foot-high (12 m) wall of water as it approached the vessel. The ship rose over each of three waves but suffered flooding of the cabins and saloons and breaking of the quarter deck. If the engraving that later appeared in *Harpers Weekly* is accurate, it is a wonder that the ship survived at all. The three coaling scows that were fueling the *La Plata* were not so lucky. One was sunk, and the other two washed onto the shore more than 1 mile away.

The *Monongahela* lost four men—three sailors who had been in her longboats and the fourth, the commodore's coxswain, who had been in the commodore's gig, which had been crushed under the *Monongahela*'s keel. In addition, three or four foolhardy sailors had literally "jumped ship" when the *Monongahela* was washed ashore and broken arms or legs. The chief boatswain of the *Monongahela* was not on the ship at the time but had his own experience with the tsunami. He had been on shore riding a pony past a local woman and her son who were gathering wood. When he saw a giant wave approaching, he told the boy to grab hold of the pony's tail and the woman to run for her life. The pony managed to outrun the tsunami and the boatswain and boy were saved, but the boy's mother was killed by the wave.

The *La Plata* survived the tsunami, as did the *Susquehanna*. The *De Soto*, though damaged, had also survived and would later rejoin the fleet after repairs in Norfolk Naval Shipyard in Virginia. The *Monongahela* was high and dry, but Admiral Palmer, commander of the squadron, decided to try and have her refloated. Several weeks later, however, the admiral died of yellow fever, not living to see the *Monongahela* successfully repaired and refloated 6 months after the tsunami. She arrived safely in New York on June 1, 1868, and would later serve as the training vessel for the US Naval Academy.

Not surprisingly, the deal to purchase the Virgin Islands and establish a US naval base was sunk by the disaster. Only 3 weeks before the earthquake and tsunami struck, the Virgin Islands had been hit by a devastating hurricane that had damaged nearly every building on the islands. Now, with the added destruction from the earthquake and tsunami, the US Senate balked and would not ratify the treaty to purchase the islands. Although the Danish government and the islanders were overwhelmingly in favor of the transfer to the United States, the tsunami delayed the ultimate acquisition of the Virgin Islands by the United States until 1917.

Ironically, the decision to not proceed with the purchase of the Virgin Islands in 1867 had some long-term repercussions. Attention was refocused toward the Hawaiian Islands and a possible Pacific naval base at Pearl Harbor. A few years before, the US, British, and French governments had recognized the harbor as a possible location for a naval base but were put off by shallow water and reefs blocking the entrance. Ultimately, the United States and Hawaii signed the Reciprocity Treaty of 1875, and it was so economically beneficial to Hawaii that when time came for its renewal 7 years later, Hawaii sweetened the deal by granting

> to the Government of the U.S. the exclusive right to enter the harbor of Pearl River, in the Island of Oahu, and to establish and maintain there a coaling and repair station for the use of vessels of the U.S. and to that end the U.S. may improve the entrance to said harbor and do all things useful to the purpose aforesaid.

Needless to say, Britain and France were upset, but the United States ignored them, with the US Navy eventually taking possession of the harbor in 1887. Since that time, the harbor has had an interesting history, but one little known fact is that it is now home to the Pacific Tsunami Warning Center (PTWC). This is particularly ironic for two reasons. First, the main reason for establishing the PTWC was because of the devastating effects of the 1946 Aleutian tsunami and so placing the warning center close to sea level within Pearl Harbor is an interesting choice due to possible exposure to tsunami inundation. Second, the refocusing of the attention of the US Navy away from the Caribbean because of an earthquake and tsunami in 1867 moved it out of the frying pan and into the fire with regard to tsunamis. In the short period between approximately 1812 and the present day, the Hawaiian Islands have experienced well over 150 tsunamis. Fortunately, only a few have been catastrophic, but the exposure is nevertheless high. On the other hand, during the same time period, the US Virgin Islands have possibly experienced up to 16 tsunamis, of which only 2, including the 1867 event, have been of any concern.

Interesting decision.

The Strange Last Voyage of the USS *Wateree*

In 1922, Captain J. K. Taussig, in command of the US Navy cruiser *Cleveland*, was delivering aid to towns and villages along the coast of Chile. The Great Atacama Earthquake, with a magnitude later estimated as high as 8.3, had struck on November 10. The massive earthquake caused notable movement of the land all the way to the seafloor and had generated giant waves. These terrible waves had surged more than 1 mile inland, devastating coastal communities. Captain Taussig had never before experienced tsunami waves, then called "tidal waves," but he had heard of them. His grandfather, Midshipman Edward Taussig, had been through an even more devastating tsunami more than 50 years earlier, some 600 miles farther north. Thus inspired, the captain decided to investigate his grandfather's experience. This is what he found out.

In 1868, two US Navy ships were involved in a terrible tsunami. One vessel, a sidewheel steam gunboat, the USS *Wateree*, had been built in 1863 for the Civil War and assigned to the Pacific Fleet. Following the Civil War, the Pacific Fleet was divided into North and South Pacific Squadrons. Assigned to the South Pacific Squadron, the *Wateree* had been patrolling the coast of South America and was in port in Arica, near Lima, Peru, on August 16. Also at Arica that day was the Peruvian gunboat, *Americana*, and the US Navy store ship, *Fredonia*, carrying $2 million worth of supplies to support the South Pacific Squadron.

At 5:05 p.m., a rumbling noise was heard on board the *Wateree*, followed by a trembling motion of the ship. A major earthquake (later estimated to have a magnitude of 8.5) was occurring, and a seaquake (the seismic energy transmitted through the water and felt onboard ships in much the same way as an earthquake is felt on land) had just struck the vessel. The captain, James Gillis, rushed on deck in time to see the entire city of Arica reduced to a "mass of ruins" in "less than a minute." Captain Gillis then gave orders to secure the gun battery and batten down the hatches. The captain and ship's doctor had then gone ashore to render aid to the victims of the earthquake when, at 5:32 p.m., the sea began to rise rapidly. Lt. Commander Steyvesant, now in command of the *Wateree*, released one of the anchors and started the engine. As the vessel swung around in the turbulent water, it took four seamen to control the helm. But the remaining anchor held, and the *Wateree* maintained at her position in the harbor. The *Fredonia*, however, could be seen near shore lying on her side with her deck facing the sea. During the turbulent withdrawal of the water, several small boats were drawn past the *Wateree* and their occupants hauled aboard. Then a man was seen drifting past in a mass of bushes and appeared to be drowning. Midshipman Edward Taussig, future grandfather of the Navy captain who would later share his story, volunteered to render aid and left in the ship's small cutter. No sooner had the cutter left the relative safety of the *Wateree* when a second wave, this one a monster perhaps 90 ft (27 m) high, surged into the bay. The officers and men of the *Fredonia* worked bravely and frantically to save their vessel, but she was taken by the wave, carried toward Alacran Island at the southern end of the harbor, and smashed to bits on the rocks by the avalanche of water.

The cutter with Midshipman Taussig was carried back by the current toward the *Wateree*, where he was thrown a mooring line. "Houses, railroad cars, men, dogs, trees, and miscellaneous debris" swirled around the vessel. Suddenly, the line holding the cutter broke and Taussig and his coxswain were adrift in the surging current. They were carried in the direction of the Peruvian warship, *Americana*. With hope of finding safe refuge on board the warship, both the officer and the sailor took oars and rowed. "I never pulled so hard in my life" Taussig would write to his parents a few days later. As they pulled near the Peruvian ship, a line was thrown to them, but at the very moment they grabbed it, the cutter was smashed against the warship and began sinking. Just as the boat filled with water, the two seamen were safely hauled aboard the *Americana*.

Meanwhile, heavy seas were now beginning to break over the *Wateree*. Both anchor chains were out to their bitter ends and straining, when suddenly they parted close to the hawse pipes. The *Wateree* was now adrift in the current and headed toward Alacran Island. It just missed the rocky island and began drifting rapidly toward shore.

Back on the *Americana*, Midshipman Taussig had stepped into a scene of utter chaos. The captain and approximately 85 men had just drowned, and the remaining crew, having broken into the liquor stores, were crazy drunk. The lieutenant left in command had tried shouting orders but finally gave up in despair and wept. Waves swept across her deck, and the masts began falling overboard. Taussig lashed himself to the shrouds and hoped for the best.

At approximately 7:20 p.m., both the *Wateree* and the *Americana* were washed ashore and grounded (Figure 7.1). The *Wateree* had lost only one crewman, the seaman in charge of the captain's gig, who had been on the beach when the largest wave struck and was carried out to sea by the current. The *Fredonia* had lost all hands except for two sailors, who somehow miraculously survived the destruction of their vessel, and the captain and paymaster, who had gone ashore to render aid in town.

The town of Arica had not fared well either. According to Rear-Admiral Turner, in command of the South Pacific Squadron, the upper part of the city had "not a single house or wall left standing" and was "a confused mass of ruins," and the lower parts of Arica were "perfectly swept clean, even the foundations, as though they had never existed."

The main job of the surviving Americans was now to render aid to the Peruvian survivors and bury the dead. The crew of the *Wateree* went about distributing supplies to the survivors of the disaster, and although the *Fredonia* had been utterly destroyed, many of her stores had fared rather well and were strewn along the beach. Surprisingly, the bureau from the paymaster's state room on a lower deck of the ship had been washed ashore without losing a single drawer. Among the supplies was an ample store of liquor, and it was said that "for three days [after the disaster] even the most humble 'cholo' [Indian] would drink nothing but champagne." Tents were made of whatever

Figure 7.1 The *Americana* was broken and half destroyed, lying very near the *Wateree*. No crew members were found. The stranded *Wateree* with the *Americana* to the left in the background are shown (the shattered hulk of the English sailboat HMS *Charnasilla* is seen in the background to the right). No crew survived.

Source: Renato Aguirre Bianchi (http://www.aricaacaballo.com/arica_territorio_andino/arica_territorio_000184.htm).

material could be scavenged from the beach, including several tents made entirely from maps of Bolivia.

A careful survey of the *Wateree* proved that she was practically intact, but because she lay 430 yards (390 m) from the sea, it would be impossible to launch her. However, the vessel stayed fully commissioned for several months following the disaster. The crew constructed heads and washing facilities on shore and started a small vegetable garden, but strict naval discipline was maintained even though the ship was high and dry. This, however, did result in some rather unusual naval practices. For example, instead of using boats to get around, mules became the most useful means of transportation. Mules were used in the local nitrate industry, and following the earthquake, many were found roaming the sand dunes. The sailors needed only to "kidnap" and press them into service with the US Navy. To facilitate the use of mules instead of boats, the animals were hitched by lanyards to a lower boom in readiness. When the captain wished to go for a canter among the dunes, the officer of the deck would pass the word to the boatswain's mate, who would call out "First Mule" or "Second Jackass" and the coxswain would slide down a line to a burro and come alongside to the ship's ladder in readiness for the commanding officer to mount his steed.

Eventually, the *Wateree* was sold at auction to a hotel company and used as an inn, with her guns being sold to the Peruvian government. An outbreak of yellow fever later turned the ship into a hospital, and then she was moved closer to the sea by another tsunami in 1877 (Figure 7.2). Later still, during the Peruvian–Chilean war in 1880, her hull was bombarded by the Chilean Navy, and her guns were used by the Peruvians to bombard the Chileans. By the turn of the century, all that was left were her gaunt ribs and boilers rising above the shifting sands. Today, all that remains are parts of her boilers adjacent to the highway along the shore north of the present town of Arica—they were moved here to be close to the highway . . . and the sea. She is getting closer to the sea at every stage of her journey. The boilers are maintained as a National Monument of Chile (Figure 7.2). There is one other little bit of her still remaining. Hidden in a hole in the ground in the location where she was deposited by the 1877 tsunami is a small piece of rusted metal. As far as we know, this is the only remaining piece of her still preserved at her final resting place after her adventures with two tsunamis in the space of 9 years.

What became of Edward Taussig, the courageous midshipman? Those qualities of intelligence and courage that he had displayed so well during the tsunami of 1868 would ultimately lead to his rising through the ranks to become Admiral Taussig.

Figure 7.2 (a) A rusty piece of metal at an archeological site in Arica—the last remaining fragment of the *Wateree* where she was deposited after being moved by the 1877 tsunami from her initial resting place farther inland following the 1868 tsunami. (b) The boilers of the *Wateree*, with the city of Arica in the background.

Source: J. Goff.

In some ways, not much has changed since the days of Taussig's Navy. Even modern ships can be trapped and destroyed in shallow water by tsunami waves. Indeed, it was almost 100 years later in 1964 that ships along the coast of Alaska felt the full force of tsunami waves and, just like Arica, the ports affected were changed forever. The only difference in this case was that we have a richer collection of survivor stories that tell us so much more about the events that took place on March 27, 1964.

8

Not Such a Good Friday for Alaska

I was standing on the lake shore, admiring the stillness that was settling in with the approaching darkness, when I thought I felt the ground move slightly. The lake's calm surface seemed to ripple, and a couple of small waves lapped against the shore. There wasn't a breath of wind, yet the air seemed to change. I wasn't alarmed—just sort of intrigued—as I returned to the cabin. I turned on the radio to get some music, but all the stations had the same thing: news about the disastrous earthquake in Alaska. Here's my recollection of the sensationalized news reported during the first few hours after the quake:

Anchorage has been leveled. . . . Valdez and Cordova have been washed away by a giant wall of water. . . . Seward is a sea of flames . . . fishing boats in Kodiak are resting on the sides of the mountains. . . . Alaska has been devastated . . . thousands are feared dead.

—Frank Baker, 2018—recollections while vacationing near Vancouver, Washington, in March 1964

Not a Tsunami, But . . .

Frank Baker experienced a seiche, sometimes called a seismic seiche, caused by the Alaska earthquake of March 27, 1964. A seiche is a bit like sloshing water around in a bathtub.

Seiches are not tsunamis but they can be generated by earthquakes, although they are usually caused when strong winds and rapid changes in atmospheric pressure push water from one end of a body of water to the other. When the wind stops, the water rebounds to the other side of the enclosed area. The water then continues to bounce back and forth for hours or even days. If an earthquake is large enough, then as its seismic waves pass through an area, they can cause this sloshing motion.

Seismic waves from the Alaska earthquake were so powerful that they caused water bodies to oscillate in many places in North America and were even observed as far away as Australia. That's pretty impressive.

Interestingly, the first published mention of this weird phenomenon that we are aware of was following the 1755 Lisbon earthquake, where numerous seiches were reported in the United Kingdom from Scottish lochs to English moats (this is further discussed in Chapter 12).

Focus

Tsunamis do not always do what people think they will do. For example, it depends where they start. This may seem logical, but it is interesting to see in practice. In the Southern Hemisphere, the powerful subduction zone just offshore from the Chilean coast has generated some huge earthquakes and tsunamis. The 1960 tsunami (see Chapter 14) was generated by a large earthquake in the southern part of Chile—it crossed the entire Pacific Ocean, focusing on and inflicting fatalities in both the Hawaiian Islands and Japan, but it did little or no damage in New Zealand directly opposite Chile across the South Pacific. However, the 1868 tsunami (see Chapter 7) was generated by a far smaller earthquake in the geographical "armpit" of South America on the Chile–Peru border in northern Chile. It is the largest distant-source tsunami to have struck New Zealand in historic time even though the country is much farther south from this point of South America on the other side of the ocean. Two things cause this focusing of tsunami energy. First, the tsunami takes a great circle route around the world—the shortest distance between two points. That is why when you fly from Europe to the northern United States and Canada, you tend to go over the Arctic; no airline is going to waste fuel traveling too far. Second, subduction earthquakes such as those that occurred in 1868 and 1960 are generated by movement of an undersea fault line, and if that line is facing in your direction, then the tsunami will come your way even if it is geographically far to the north, for example.

In the Northern Hemisphere is the Aleutian Trench. This runs from the geographical "armpit" of Alaska/Canada to Russia, crossing the International Date Line. Although it is underwater, its path can be tracked by following the Aleutian Islands, which are a volcanic island arc—a long chain of active volcanoes that sit "behind" a convergent tectonic plate boundary. If an earthquake happens approximately one-third of the way along the trench from the east (near the beginning of all of the smaller Aleutian Islands), such as the devastating 1946 event, then it will be a major threat to the Hawaiian Islands and Antarctica (as discussed in Chapter 1). On the other hand, tsunamis generated by earthquakes toward the eastern part around the Gulf of Alaska, such as the 1964 event, pose a greater threat to the population centers of Alaska and also to the west coast of Canada and the United States (Figure 8.1-see color plate section).

1964

At 5:36 p.m. Alaska Standard Time on Good Friday, March 27, 1964, the largest earthquake ever recorded in North America—a colossal 9.2 magnitude quake—struck Alaska, rocking the southern part of the state for 4 or 5 minutes. The quake was centered near the eastern shore of Unakwik Inlet in northern Prince William Sound. It was a shallow earthquake, which basically means there is more shaking and a greater likelihood that a tsunami will be generated. The quake was accompanied by vertical movement over an area of some 200,000 square miles (>500,000 km^2) and ranged from subsidence of 7.5 ft (2 m) to uplift as great as 38 ft (11.5 m), including the islands and mainland of Prince William Sound. It was so large that it caused the entire Earth to ring like a bell—vibrations that were among the first of their kind ever recorded by modern instruments. The earthquake was followed by a seemingly endless number of lesser aftershocks, and it caused hundreds of damaging landslides both on land and below the sea.

In Alaska, tsunamis immediately began to wreak havoc. In fact, tsunamis from two different sources struck some areas. In many harbors near the epicenter, the seismic waves caused submarine landslides that themselves generated tsunamis. Some 20 highly destructive local tsunamis caused by these landslides would have devastating effects on the communities at Valdez, Whittier, and Seward. Meanwhile, the main movement of the seafloor had generated a major tsunami that would be felt all around the Pacific basin. Tsunamis caused casualties and damage from the Kodiak Islands in the eastern Aleutians all the way down the west coast to northern California. Altogether, the earthquake and subsequent tsunamis caused 131 fatalities and an estimated $2.5 billion in property losses (in 2020 dollars).

Of the 131 people killed, between 119 and 122 died in tsunamis. In the village of Chenega, on Chenega Island, one-third of the population was killed and the only building that remained was the schoolhouse, built 100 ft (30 m) above sea level. Ultimately, the survivors moved to Evans Island some 10 miles to the south. Residents of several other towns and villages also moved to safer ground after the tsunamis, including Seward and Valdez. Farther afield, 13 people died in California and 5 in Oregon, but amazingly no one was killed along Canada's Pacific coast, although the tsunami hit Port Alberni on Vancouver Island's west coast twice, washing away 55 homes.

Figure 8.2 The Old Valdez pier and docks prior to the 1964 tsunami.
Source: Thalma Barnum Collection, PTM.

One of the highly destructive local tsunamis rose to a height of 219 ft (67 m) in Shoup Bay, Valdez Inlet. Five miles (8 km) to the east lies the port of Valdez.

Valdez

Only 40 miles from the epicenter of the earthquake lay Port Valdez, the northern-most ice free port in Alaska and as such an important terminal for transport to the interior of the state (Figure 8.2). It is a former gold rush town situated at the head of Valdez Arm, a steep-walled fjord in the northeastern corner of Prince William Sound. There is not much room to build ports in those fjords, and so the town of Valdez had been built on the unconsolidated sand and gravel of a delta near the head of the valley. It was devastated by the tsunami, but like so many of the coastal towns in the area, it is replete with remarkable survivor stories.

March 27 was a gloomy spring day with ice still on the roads. Just after 4:00 p.m., the 400-ft freighter, SS *Chena*, arrived from Seward and tied up at the dock of the freight terminal. Local longshoremen went onboard to transfer cargo, and as usual a crowd of adults and children gathered on the dock to watch. Among the children gathered there was 13-year-old Fred Christoffersen and his buddy Danny Feeks. Christoffersen stated,

> We went down there and one of the cooks was pitching apples and oranges out the porthole of the ship so we stuffed our pockets full of fresh oranges and apples, and I told Danny . . . I says, "You know, I need to go to the bathroom." He said, "Well, let's go to the warehouse. You can go in the corner."

The two boys walked off the pier and down toward the warehouse. "We had just stepped off the wood portion of the dock onto the dirt road and I threw my apple core over the side and it seems like as soon as that apple core hit all hell broke loose." Danny immediately shouted,

> Earthquake . . . run. So, we took off running . . . and I kept turning around to watch . . . see what was going on. And, uh . . . it sounded like a bomb had gone off. I turned around . . . and the ship was . . . going up in the air on the first wave.

It was already near low tide when the earthquake struck, but then almost immediately the water withdrew even farther from the beaches, leaving much of the delta on which Valdez was built exposed above sea level. This suddenly increased pressure on the sediment, which along with the ground vibrating from the quake caused a section of the delta some 4,000 ft long (~1,220 m) by 600 ft wide (~185 m) to slide into the sea, carrying with it the dock and portions of the town of Valdez.

On board the *Chena*, Captain Stewart, master of the vessel, was sitting in the galley dining room when the earthquake struck: "I made it to the bridge [three decks up] by climbing a vertical ladder. God knows how I got there." The *Chena* then rolled and pitched violently to port. According to the Captain, "The Chena raised about 30 ft on an oncoming wave. The whole ship lifted and heeled to port about 50°." Suddenly, the ship's propeller was up in the air, and then the pier began to slide underneath the ship and the crewmen of the *Chena* watched in horror as the ship came down on top of the warehouse and dock. In Captain Stewart's words,

> Then it [his ship] was slammed down heavily on the spot where the docks had disintegrated moments before. I saw people running—with no place to run to. It was just ghastly. They were just engulfed by buildings, water, mud, and everything. The *Chena* dropped where the people had been. That is what has kept me awake for days.

Not a single one of the 28 people on the dock survived.

Fred and Danny were frantically running down the road away from the collapsing docks as the road twisted from side to side. Fred said,

> As the road tilted to the right, this semi [truck] started across the road in front of me so I had to . . . had to stop and wait for it to go on by on the downhill side so I could cross on the uphill side.

The cannery at Valdez now collapsed into the bay. Captain Stewart knew that he had to get his ship away from shore and out into open water:

> I signaled to the engine room for power and got it very rapidly. In about four minutes, I would guess, we were moving appreciably, scraping on and off the mud as the waves went up and down. People ashore said they saw us slide sideways off a mat of willow trees.

Meanwhile, the wave was bouncing back off the shore and washing back toward the bay:

> A big gush of water came off the beach, hit the bow, and swung her . . . out. If that hadn't happened, we would have stayed there with the bow jammed in a mud bank and provided a new dock for the town of Valdez. The bow pushed through the wreckage of the cannery. We went out into the bay.

The captain had saved his ship, but many members of his crew and longshoremen had lost their lives.

Back to Fred and Danny. They had continued running up the road toward the center of Valdez. They had just gotten as far as the town pub, ironically named the Old Village Morgue, when the walls started to collapse. They continued running, dodging falling power lines and sparking wires and jumping a 6-ft crevasse that suddenly opened up right in front of them. They kept running uphill until an acquaintance drove by in a truck and told them to jump in. Danny said, "No, I gotta go home and see if my mom's okay." He said, 'No, get in. Your mom's okay.' So, anyway, being a kid, I took the adult's word and jumped in." They drove uphill, eventually meeting up with family friends who took them to stay at their place, safely outside the disaster area. Later, Danny got a ride to his aunt and uncle's place in Glennallen, some 120 miles from Valdez, but Fred stayed and continued to be worried about his parents.

When the earthquake had struck, Fred's mother, Thelma Christoffersen Barnum, had been at home and thought that Russia had dropped a bomb on

Anchorage. She crawled around on the floor until the most violent shaking had finally stopped. Then she instantly thought of her son:

> I knew my oldest son, Fred, had gone to the dock because the ship was in. I didn't want to leave without finding my children. And I was trying to run away . . . but my husband pushed me in the car against my will, and we left town.

The next morning, her husband got a ride down to Valdez and found their daughter, but Fred was still missing. Two days after the earthquake, Thelma was still grieving and hoping for a miracle when after Sunday school, as Fred recalls, "I walked in and mom looked at me and she was pure white and she says, 'Frederick, is that you?' Yeah, mom. I'm here."

A happy ending—one that captures the chaos and disorientation of a community in crisis. Valdez was fortunate, however, because the first tsunami wave to reach it had in all likelihood been caused by the submarine slide of the delta front. The wave was estimated to be 30–40 ft (9–12 m) high, but its impact would not be limited to Valdez. A part of the tsunami wave originating at the port had been propagated westward down the bay, surging up and over living spruce trees more than 100 ft above sea level, leaving some as large as 2 ft in diameter broken and splintered. Farther down the fjord at the Cliff Mine, the tsunami reached more than 170 ft above sea level and then surged out through the Valdez Narrows and was described as a black wave with mud, rocks, and other debris.

This is very reminiscent of the landslide-generated tsunamis in the remote fjords of Alaska such as Lituya Bay (see Chapter 5) and Taan Fjord (briefly mentioned in Chapter 6), but it was happening in an area where people had decided to settle.

No one in the town of Valdez could see what was going on past the dock out in the bay. The turbulence of the first two waves had apparently created a mist or haze that obscured their view beyond the shoreline. They couldn't see that 68 of the 70 boats in the harbor were almost immediately destroyed. Fortunately, no one was on board any of them at the time, but in total, 32 people had lost their lives to the tsunami in Valdez.

As is the case with many modern tsunamis, fire broke out. The sky was lit up as the tanks at the Union Oil Company, ruptured by either the earthquake or the tsunami, leaked fuel that then caught fire. By 10:30 p.m., the entire waterfront area of Valdez was in flames. Smaller tsunami waves from the main earthquake-generated tsunami washed ashore at 11:45 p.m. and 1:45 a.m., but they failed to put out the fires, which would burn for another 2 weeks.

Following the 1964 earthquake and tsunami and the horrific loss of life at Valdez, the decision was made to move the entire community to a safer area—one that was more stable and had some natural protection from tsunami waves.

Seward

At the head of Resurrection Bay 120 miles (190 km) southwest of Valdez and on the southeast coast of the Kenai Peninsula sits Seward, Alaska. Resurrection Bay is another steep-sided fjord, but it faces directly south to the open ocean some 20 miles (~30 km) away. Like Valdez, the deep fjord allowed large ships to enter, the port was sheltered from the bad weather of the open coast, and the river flowing in at its head provided flat land to build on . . . perfect?

Seward was a town of approximately 1,700 people in 1964; it was an important marine and railroad terminal. When the earthquake struck, the coastal tanker, *Alaska Standard*, was tied at the Standard Oil Company's fuel dock by seven mooring lines, and five large fuel hoses were in use loading gasoline, stove oil, and diesel fuel. Seaman Theodore (Ted) Pederson was on so-called hose watch on the dock. Within 30 seconds of the onset of the quake, the waterfront began to shake violently. The tanker first bucked and then slammed against the pier. Ted began to run up the dock toward shore as fast as he could while the wharf pilings began shooting into the air around him and the 200-ft-long warehouse next to the dock began to sink down as entire sections of the Seward waterfront started sliding into the bay. Some 14 storage tanks ruptured, and their fuel began to spread across the water. Then the *Alaska Standard* heeled sharply away from the pier, breaking the fuel hose connections and causing "geysers of oil to shoot skyward." Ted had run almost 100 ft (30 m) up the dock when the fuel caught fire and the tank farm blew up in "a big ball of fire." Next, according to the ship's master, Captain Solibakke, the tanker dropped vertically 20–30 feet, hitting bottom, and then "jumped straight up in the air," landing on the dock—the same dock Ted was standing on. Ted fell into the water and looked up just in time to see a huge wave, filled with debris from the dock, coming down on top of him. He was struck on the head and lost consciousness.

Just like at Valdez, this was another local tsunami, generated by the sliding of the Seward waterfront into the bay creating a 30-ft wave. The tsunami wave, showing no favoritism, now carried the flaming oil to the nearby Texaco petroleum facilities and, just like Standard Oil, set them on fire too. It was truly a scene from hell, and the fire caused by the tsunami was now after Captain

Solibakke's ship. The *Alaska Standard* had been washed away from the burning dock by the wave, but the ship, with its cargo of gasoline and oil, was almost surrounded by flames on the surface of the water. With the ship now under power, Captain Solibakke managed to skirt the flaming water while heading his vessel toward the entrance of the bay, as his crew fought small fires that had already broken out aboard ship.

Meanwhile, Ted had regained consciousness. He was surrounded by debris and had a broken left leg, but he was alive and was somehow onboard his ship, the *Alaska Standard*. Ted had miraculously been washed onto his own ship by the tsunami wave and had come to rest on a catwalk 8 ft above the ship's deck. A lucky man.

But what about those onshore?

For many people faced with a tsunami, there are two main responses. First, the immediate urge is to stand there like a deer in the headlights as your brain tries to process a phenomenon that is simply way beyond your experience and so you take no action and simply watch it. Second, you jump in your car and try to get away as fast as you can.

It was just another day for brothers Bob and Mack Eads, except "We noticed there weren't any dogs and cats around; there weren't any birds. . . . We were wondering where all the seagulls were."

This interesting observation is regularly commented on by tsunami survivors. In the 2004 Indian Ocean tsunami, survivors reported the movement of wildlife inland approximately 30 minutes before the tsunami struck southern Sri Lanka. It is difficult for scientists to study such a response: Is it real or just a perception imprinted after the event? Because we cannot predict when such catastrophes are going to happen, it is difficult to monitor the wildlife and so such observations tend to be viewed with some skepticism by scientists. Nevertheless, it is a common theme—so perhaps there are environmental signals about the impending disaster that we are not picking up.

The brothers were finishing their day's work at their shipyard business at Lowell Point, just down Resurrection Bay from Seward, when the earthquake struck. They saw the Standard Oil tank yard burst into flames and then the 30-ft, fire-covered wave heading right toward them. They had never seen anything like this and so, far from running away, Bob grabbed his movie camera and started to take pictures. They believed they had time, but tsunamis move faster than normal waves. Fortunately for them, their brother-in-law, Carl, was from Hilo, Hawaii, and he knew about tsunamis. Carl and Bob jumped in the car and took off ahead of Mack, who quickly got in his truck. They had only gone a few feet when the wave caught them, flipping the truck end over end and slamming the car into the trees. Bob and Carl managed to get out of the smashed car and climb up on a boat in dry dock, but Mack was trapped in the truck underwater: "I was drowning. I was getting a lot of mud and water and dirt into my lungs and into my throat." His truck continued to be washed all around by the wave until finally it came to rest on its side. "My driver's window was open and I just . . . propelled myself out of that pickup into the water." Just as he climbed out of the truck, he was struck by a large log:

> And it hit me . . . the full length of my body . . . paralyzed that whole side. I was still in the water and the wave started going back. I was going back toward the ocean. I was just sailing along and I grabbed an Alder and hung on until the water drained away and there I was . . . I was just like a beached sea lion . . . laying there.

From the relative safety of the boat up in dry dock, Bob and Carl

> started looking around to see what was going on. That's when we heard . . . we thought it was a sea lion groaning there on the road. It was Mack trying to puke the mud out of his stomach. We went down there and found him . . . got him back up on high ground.

Linda McRae and her family had a similar experience, although this time it involved a house. Like cars, houses can also float. They had all jumped in their car following the earthquake, and as they drove down the road to pick up more family members in Seward, they were amazed to see "the radio station . . . and the trailer park was floating in the water."

No one was home—they had already fled to higher ground—but Linda and her family went into the empty house and were "there just a few minutes when . . . my father is at the front door yelling, 'There's a tidal wave coming. We have to get to the top of the roof.'" They frantically clambered over oil drums onto a shed and finally made it to the flat roof of the garage, some 6 ft away from the main house. "We could see that the water had pulled back and there was a large, dark wave headed down the bay to pretty much exactly where we were." Now they realized that they had to get even higher to the roof of the main house. "People just started jumping . . . my mom, sister, little brother, sister-in-law. . . . I remember that I had the baby zipped in a jacket, we all went across." They were all holding hands now, lying at the peak of the slanted roof and watching as the homes below them "instantly exploded. And we went from the ground up to the very top of the trees . . . spinning round and round. And the baby just slept through it all." They continued to be washed inland until finally "what remained of the house . . . slammed in a strand of trees."

They all thought that it was over, although they could still see "huge fires burning. And then we start hearing this loud whistle . . . like wind coming . . . but these were more waves." Gradually, it began to calm down, so her father and brother climbed down and managed to build a little fire lower down the hill to alert them if the water came up again. And then

that sound came . . . the rushing water . . . it continued to come up and the fire went out and the waters still rising . . . and that water crossed my ankles. And my brother's begging his wife to take the baby and climb up the trees . . . and she wouldn't do it. And then it started to go down . . . and that was the last wave.

When eventually they all walked out together and arrived at the "site accounting for people," they found their names on the "assumed dead list. . . . There aren't really words to describe how you feel when you see 'assumed dead' and are able to say, 'No, we're fine" (Figure 8.3).

Not everyone would be so lucky; a total of 12 people died at Seward as a result of the tsunami, and there were many close calls. The Standard Oil and

Figure 8.3 Linda McRae (far right) with her family climbing down from the house they floated on into the "strand of trees."
Source: Linda McRae McSwain Collection, PTM.

Texaco tank farms would burn for 2 more days before finally being brought under control.

Kodiak

Some 300 miles from Prince William Sound lies Kodiak Island, far enough from the earthquake epicenter to be spared the devastating local tsunamis generated by the intense seismic activity. But the Kodiak Island group was not outside the range of the main tsunami waves, and the islands that comprise it would be the only ones outside Prince William Sound to experience heavy damage. In fact, Kodiak Island actually shielded the Aleutian Islands and much of the Alaskan Peninsula from the main tsunami.

The city of Kodiak and surrounding areas are a major population center and in 1964 counted some 4,200 residents. Following the earthquake, the mayor of Kodiak met with the police chief and city manager to discuss what action to take. They were aware of the potential danger of a tsunami, but they took no action until an isolated native village on the opposite end of the island was struck by tsunami waves. Finally, they decided that they needed to issue some kind of warning, but the earthquake had left the area without much in the way of electricity or telephone service, and the Civil Defense siren was out of operation. With no plan for such a situation, the mayor decided to have the fire trucks turn on their sirens. Many residents responded by going into the streets to see why the sirens were being sounded, and police herded them up nearby Pillar Mountain toward safety.

Many people ignored the police, and ironically, many of the casualties would be local fishermen who had listened but did not sufficiently understand the tsunami danger. When they heard the warning, many of them went down to the harbor to try to save their boats. Fortunately, the first wave was a gentle flood followed by a gradual ebb, alerting many residents to the danger—they then fled to safety. But the second wave came crashing in as a 30-ft wall of water, washing 100-ton fishing boats over the breakwater and as far as three blocks into town. Although several fishermen rode out the waves, six died trying to rescue their vessels.

Many of the remote native villages are extremely vulnerable to tsunamis, being constructed along narrow strips of land just above sea level, but there is surprisingly little native lore about tsunamis. Possibly because in the past few had survived? There is, however, what might be described as a "tsunami saint" of the Russian Orthodox Church in the village of Ouzinkie on Spruce Island, one of the small eastern islands of the Kodiak Island archipelago (Figure 8.4).

Figure 8.4 Tsunami Saint in the Russian Orthodox Church, Ouzinkie, Spruce Island, Alaska.
Source: Walter Dudley Collection, PTM.

In August 1970, Father Herman (1751–1836) was canonized at Kodiak. He is now a venerated, if rather colorful, figure. The monk was sent to Alaska by none other than Catherine the Great and settled on Spruce Island near Kodiak, where he was believed to have lived in a cave and worn a reindeer jumper. He slept on a hard bench with bricks for a pillow and no blanket. He ate little and was said to have worn 15-pound chain fetters around his ankles. Among his saintly acts, one in particular is of interest. According to legend, a tsunami was approaching the village of Elovio Ostrof when the inhabitants ran to the good father for help. He is said to have taken an "image of the Blessed Virgin" to the shore and placed it on the beach. Here, he knelt down and after a short prayer turned to the assembled people and said, "Do not be afraid, the water will not rise beyond the spot where the holy image stands." And apparently the future Saint's prophecy was fulfilled.

This is reminiscent of the story of the 11th-century English (he was also king of Denmark and Norway) King Canute and the tide, an apocryphal anecdote illustrating his humility. He sat on his throne on the beach and showed that he actually had no control over the incoming tide, explaining that secular power is vain compared to the supreme power of God. Clearly, he was in the wrong business.

Perhaps the saint was on duty again in 1964. Gene Anderson was born and raised in the tiny Native American village of Ouzinkie on Spruce Island, the same island where St. Herman had lived. He was at home having dinner with his family when the earthquake struck, and his wise grandfather said, "We're going to have a tidal wave." Like other fisherman, his first thought was to save their boat, "so . . . I went down to the dock down there . . . and all of a sudden you could see the water start coming in." A man on another boat yelled, "Get out of here. The tides coming up." So Gene jumped on his family's boat, untied the mooring lines, and was pulling out by the backwash and then the water flooded "over the dock seven feet." The next wave flipped his boat back over the dock "and I took off straight for deep water," where he came alongside a larger boat and a crewman yelled "Better get aboard here . . . and I jumped on the bigger boat." That boat had a radio, and they could hear the transmission from another boat farther out. "We could hear the skipper saying, 'Hang on, here it comes!' "

Then the radio went ominously silent.

Throughout the night, they continued to hear radio messages about additional tsunami waves and they "kept saying, 'You guys out there, you know . . . just don't come any closer . . . stay away. Stay away from Kodiak.' " Finally at approximately 1:00 a.m., they heard "I think it's safe for you guys to come in now" and they headed slowly back to Spruce Island. When they arrived at Ouzinkie, "I couldn't believe what I seen . . . boats up against the hill, boats sunk, lots of houses [destroyed]."

Reflecting on his good luck, Gene said, "I was thinking of saving a boat, which was replaceable. . . . People got lost on account of saving a silly boat. You can get another boat, but you can't get another body. . . . It's best to stay by your people." Their home and their boat had been destroyed, but he said their greatest loss was their family photos. They couldn't be replaced, but Gene and his family had all survived. Did he believe they had been saved by Father Herman, the Tsunami Saint? Well, no, Gene was a Baptist!

Around the city of Kodiak, a total of 15 people were killed, but the tsunamis weren't finished. Not only had waves from the main earthquake-generated event headed west toward Kodiak but also they had spread south along the west coast of Canada and the United States.

Canada

Being farther away from the origin of the tsunami, there was more opportunity for a warning in Canada. Rather ironically, however, the Canadian authorities had withdrawn from the tsunami warning system just the previous summer. The earthquake had been felt in western Canada, but there was no significant damage. All the devastation in Canada would be from the tsunami, which would strike near high tide in British Columbia.

It was along the southwestern coast of Vancouver Island where the tsunami did the most damage. Of the 20 homes in the Hesquiat village at the head of Hot Springs Cove, 18 were washed off their foundations and carried out into the inlet. Ironically, the largest wave heights would not be measured along the open Pacific coast of British Columbia but instead up Barkley Sound at the head of narrow Alberni Inlet nearly 40 miles inland. Here lay the twin cities of Alberni and Port Alberni, an industrial center known for plywood, pulp, and paper products. The dimensions of Alberni Inlet are such that its normal period of oscillation (seiche) is very close to the dominant period of tsunami waves reaching the continental shelf off Vancouver Island. Resonance between the period of seiche and that of the tsunami can produce wave heights at the head of the inlet that are two or three times greater than those along the open coastline outside the inlet.

But fortune would smile on Alberni and Port Alberni. As the tsunami reached the mouth of the inlet, an alert lighthouse keeper telephoned ahead to the twin cities giving them a 10-minute warning while the waves traveled the 40 miles up the inlet. In another bit of good luck, the first wave was rather small and gentle, serving as a warning to many residents. The Royal Canadian Mounted Police then went from house to house warning about the potential danger of the tsunami. An hour and a half later, a second, much larger wave struck. Many houses along the northeast bank of the river were washed more than half a mile upriver at speeds estimated by observers at more than 20 miles per hour. Logs and lumber piled up for shipment were turned into water-borne battering rams, and fishing boats, torn from their moorings, further added to the mass of floating debris. The water flooding into town and seeped into underground storage tanks at service stations, sending gasoline flowing into the streets. Fires quickly broke out as electrical short circuits ignited the gasoline. The highest wave had reached nearly 21 ft (~6.5 m) above sea level. Although most residents were asleep when the lighthouse keeper sent his warning, miraculously there was no loss of life or even serious injury. The area Civil Defense commander later reported, "I am unable to account for the lack of casualties."

West Coast of the United States

Continuing on its path of destruction, the waves next assaulted the coasts of Washington and Oregon state. At Ocean City, Oregon, a park ranger reported, "It came over the dunes shooting 5 to 6 feet high, tossing logs around like match sticks." At the Copalis River in Washington state, Leonard Hulbert had stopped his car on the bridge to watch driftwood piling up against one of the supports. The bridge suddenly collapsed, plunging him into the river inside his automobile. Hulbert suffered a broken arm but managed to open his car door and swim to safety. The swirling water tore two spans from the bridge.

Farther down the coast at Depoe Bay near Newport, Oregon, a family of six were camping at Beverly State Park when the tsunami caught them in their sleeping bags. Although the parents survived, their four children were washed out to sea and drowned. Much of the damage in Oregon occurred away from the oceanfront along estuary channels, where homes, businesses, and bridges were destroyed or severely damaged. The size, shape, and depth of these estuaries apparently were the main factors in determining whether the tsunami was dissipated near their mouths or propagated upstream as dangerous waves.

The earthquake in Alaska had revealed weaknesses in the tsunami warning system, but the response to the warning in California showed even larger problems. At the state capitol in Sacramento, the California Disaster Office (CDO) had received the first tsunami warning message at 9:36 p.m. local time, but it did not send any information to coastal areas. It was only after the second and third messages were received that an alert was finally sent out, but not by the CDO. Bizarrely, the tsunami alert message was sent out by the California State Department of Justice as an "all-points bulletin" to sheriffs, chiefs of police, and Civil Defense directors in coastal areas.

Crescent City, California, lies just south of the Oregon border. A coastal town named for its crescent-shaped bay, it had a history of being susceptible to tsunamis. The county sheriff's department received the alert at 11:08 p.m., but coastal areas to the north had not reported any destructive tsunami waves, and not understanding that Crescent City might respond differently to tsunami waves, the department delayed taking action. It was only after the first tsunami wave struck, rising 14 ft (~4 m) above sea level, that the sheriff's department finally began to react, sending sheriff's deputies and local police door-to-door in the waterfront areas to warn residents. But many locals, remembering only small waves from the 1960 Chilean tsunami, viewed it as a false alarm. This assumption seemed to be confirmed when a second wave 6 ft high (<2 m) came rushing in at 12:20 a.m. By now, some residents, as well as curious tourists, had returned to shoreline areas to "watch the tsunami," a common occurrence even today—for example, people in Sydney, Australia, went to the beach to "swim" with the 2010 Chilean tsunami. Police tried to keep the tourists out in order to prevent possible looting, but they continued to allow residents and business owners into the inundation zone so they could begin cleaning up after the earlier waves. But much worse was to follow.

At 1:00 a.m., a third wave crested at 16 ft (~5 m) and was followed by an exceptionally large withdrawal, which should have served as a warning. The interval between the first three waves had been approximately 40 minutes. If this held up, the next wave would arrive at approximately 1:40 a.m., just after high tide. No one in Crescent City knew that two mountains on the Pacific Ocean floor were already conspiring to destroy Crescent City by causing the tsunami waves from Alaska to bend toward the town. Cobb Seamount, 400 miles to the northwest, and another seamount more than 100 miles west–northwest of the city were slowing the tsunami waves in such a way that the energy was being focused on Crescent City. The fourth wave began right on schedule at 1:40 a.m., continuing until 2:00 a.m., when it peaked as a mighty 21-ft (6.5-m) surge. It moved through the coastal area of Crescent City as two deadly wedges, which then joined up and roared across Highway 101 toward the Long Branch Tavern. The owner was celebrating his birthday that night and serving a round of beers when the tsunami struck. As the wave surged into the tavern, the party climbed onto pieces of furniture, trying to stay above the water until there was barely room under the ceiling to breath. The water calmed as the wave crested, and two men swam toward shore to get a boat while the rest of the group climbed up onto the roof. The men quickly returned with the boat, picked up the party on the roof, and began to row toward dry land. They were almost to shore when the tsunami began to withdraw. The boat was sucked into Elk Creek and pulled toward the bay. As the boat was smashed against the steel grating of the highway bridge, one of the men managed to grab hold and pull himself to safety, but the rest were thrown into the swirling water. Only one was a strong enough swimmer to survive; the other five were lost to the tsunami. In total, Crescent City would suffer 10 deaths from the tsunami that night, with 29 city blocks destroyed or badly damaged. The tsunami damage and death toll at Crescent City alone exceeded the combined total of all previous tsunamis on record for the entire mainland US coast.

But it could have been much worse. When the tsunami advisory was received in San Francisco, attempts were made to evacuate coastal areas. And although some 2,500 people were evacuated, an estimated 10,000 people jammed the beaches in hope of "watching the tsunami waves." In San Diego, attempts to evacuate the beaches were rendered useless by curious onlookers, and Los Angeles County made no attempt whatsoever to evacuate the waterfront. If large waves had struck these areas, the casualty lists could have been truly horrendous.

The 1964 Good Friday earthquake and tsunami had resulted in 131 fatalities. It was the first known earthquake with an epicenter on land to cause a major destructive tsunami or, more accurately, many different tsunamis, because there were some 20 local ones. The death toll could have been much higher, but because it occurred on Good Friday, this may have saved many lives since schools were closed and business areas uncrowded. Of the tsunami death toll of approximately 119, some 82 were victims of local tsunamis caused by submarine landslides set off by the giant earthquake. They happened so quickly that there was simply no time for a warning, hence the many deaths in Valdez and Seward. In Alaska, the tsunami had struck at low tide—the inundation by waves would have been far greater at high tide. But as weird as it may sound, it was low tide that played a role in producing the local tsunamis that resulted in the majority of deaths. During low tide, with the sea withdrawn, the bearing pressure on the coastal sediments is increased, and this increases the likelihood of landslides. In other words, some of the disastrous local tsunamis might not have occurred had the tide been higher.

Following the 1964 Alaska earthquake and tsunamis, the United States established an additional tsunami warning center in 1967, with its headquarters in Palmer, Alaska. Now known as the National Tsunami Warning Center, it forms part of an international tsunami warning system that serves as the operations center for all coastal regions of Canada and the United States, except Hawaii, the Caribbean, and the Gulf of Mexico.

Progress was being made, but as usual, only after a disaster.

Afterword

I (WD) interviewed Thelma Barnum and her son, Fred Christoffersen, together at her home in new Valdez. Fred was a real bear of a man—tough, big, and strong. Although it had been some 35 years since the event, he related the story of his experience during the tsunami carefully and calmly. That is until he began talking about finally being reunited with his mother. It was obvious that Fred was thinking about how she must have suffered for those 3 days after the tsunami, believing her son had been killed. At that point in the interview, he looked across at his mother and finally broke down in tears. A big man with a big heart.

While conducting interviews on Kodiak Island, I heard numerous stories of cattle fleeing uphill following the earthquake. But I also collected reports of numerous cattle in coastal pastures drowned by the tsunami waves. Where did the truth lie? Were some cattle smarter than others, or was I recording planted memories of what people heard and thought had happened? Finally, after numerous additional interviews with eyewitnesses, I concluded that cattle did run uphill following the earthquake but that many cattle were killed by the tsunami. Ken Lester had reported seeing many dead cows, some "clear up in the trees." How could both be true? One of our final interviews on Kodiak Island was with Mr. Duncan Field, a local attorney and also a rancher who maintained numerous cattle. I laid out my cattle problem: "I've heard that the cows saved themselves by running up the hill, but that many cattle were also killed by the tsunami waves." Duncan smiled and responded,

> Oh, certainly. Yeah, when the earthquake came, the cows just took off, just ran. And what happened, see, there was a significant time lag from that earthquake to when the tidal wave came in. And so they came back down in the flat areas and they were grazing.

It seems that the cattle felt the seismic waves and responded by heading to higher ground—a perfect tsunami survivor strategy. However, they didn't know about the hour-long tsunami travel time from Prince William Sound to Kodiak Island, so they returned to the shoreline grazing areas before the tsunami struck and were killed by the waves.

Bizarrely, exactly 25 years later on Good Friday, 1989, the enormous oil tanker, *Exxon Valdez*, would run aground on Bligh Reef in Prince William Sound. Some 37,000 metric tons of crude oil would spill from the ship, creating an environmental catastrophe from which some areas have never recovered. So, did the name Valdez have bad karma, or was it Good Friday, or maybe Bligh Reef, named after Captain Bligh of *Mutiny on the Bounty* fame? Interestingly, a bar in Valdez where, according to legend, the captain of the *Exxon Valdez* had one too many, has a drink named in remembrance of the oil spill. The drink is made with Tanqueray gin poured over ice cubes and called "Tanker-ay on the rocks."

Strange but true. And on that note, that is exactly where we go next. It is hardly surprising that as tsunami researchers we have come across a few bizarre stories. Chapter 9 dips into this pool of stories. There are many more where those came from.

9
Strange, But True

> The ship Strathblane, bound from Honolulu, has been lost at sea. She was struck by an asteroid and sank. The captain and five of the crew went down with her.
>
> —*The Western Australian* **(1891)**

It is difficult to state that any of the events discussed in this chapter necessarily changed the world, but they are interesting nonetheless and should change either how we search for past tsunamis or how we define them. There are some instances in which a tsunami-like disaster happened but it does not fit within the strict scientific definition of the word, or something happened that is classified as a tsunami even though we don't really know. This is where we start to bend the rules a little.

There are many examples, but it is hoped that the few presented in this chapter will be food for thought.

A Whale of a Time

During the 2004 Indian Ocean tsunami, it was reported that a humpback dolphin (*Sousa chinensis*) was washed 0.9 miles (1.4 km) inland in Thailand. Fortunately for the dolphin, there was a happy ending. It was rescued and returned to the sea alive. This also meant that it didn't become a later conundrum for scientists trying to determine what a dead dolphin was doing that far inland. An equally fortunate marine species, a shark, ended up in a hotel swimming pool in Thailand—however, its fate is unclear.

It is not really surprising that large marine species get washed inland by tsunamis; everything else does. But interestingly, vertebrate remains are rare finds as evidence of past tsunamis, although we know that many people die and animals are killed. For example, there are few, if any, reliable examples of human skeletal remains found that can be directly linked to tsunamis (see

Chapter 4) other than in the most recent events, in which death and destruction form the bulk of any media coverage. However, an interesting example from an early tsunami does include an array of skeletal remains. The historically documented 1293 Kamakura tsunami in Japan is at the cusp of written history, with accounts of the event recorded in ancient documents such as the *Kamakura Oh Nikki* and *Azuma Kagami*. These state that between 20,000 and 30,000 people died, with the bodies of humans, cows, horses, fish, dolphins, and whales filling the roads. A mixed mass burial confirms these historical accounts and goes to show that there can be a real variety of species carried inland from the sea as well as those killed on land.

If we take this information with us further back in time, then an unusual geological find starts to make us think that a tsunami was probably responsible. At the start of the 20th century, a gentleman scientist working in Wellington, New Zealand, reported "a number of cetacean skeletons, crumbling to powder, yet preserved in form in dry sand, lie at heights up to 147 ft above high water mark. One, stretching over 60 ft, is half a mile inland." This report raises some interesting points. These whale skeletons are intact, they are not disarticulated bones, and they do not show any signs of having been cut up to eat. This not only makes it impossible for humans to have carried them nearly 150 ft (45 m) uphill (why bother anyway?) but also the lack of cutting up by humans of such a welcome supply of free food means that these whales likely arrived at their resting place before humans were living in the area.

They were found preserved in sand dunes sitting on top of a cliff. These sands arrived at the top of the cliff approximately 6,000 years ago when rising sea level pushed a large amount of sand ahead of it and it was subsequently blown uphill and deposited by the wind. Once that process was finished, there was no more excess sand around the coast to blow inland, so we have a rough idea of when the whales got there since they were encased within these sands.

If this was the result of a (very large) tsunami—which seems to be the only plausible explanation—why has no one ever found evidence for this event before? The answer is that the whales and associated sand might have been the only obvious evidence of the event, and the site has been built on—a beautiful, cliff-top subdivision comprising gorgeous houses with stunning views.

If modern scientists had not read these earlier reports from well over 100 years ago, then this mystery would not exist. And obviously the reading of these reports happened after the houses were built. Hey ho.

A century ago, it was observed that "if the whales were washed there then one of two hypotheses may be accepted: that a tidal wave [*sic*] carried the carcasses far inshore, or that the land has been raised with the embedded carcasses in it." This is a remarkably astute statement bearing in mind that it was made nearly 100 years before the modern era of research into prehistoric tsunamis (tidal waves) began in the late 1980s. It is also remarkable because we now know that part of the explanation does indeed relate to the land being uplifted by past earthquakes, but this only represents a minor component of the 150 ft (45 m) of land above sea level. So, could the second part of the explanation truly be a tsunami?

To start with, it is useful to know that this location is on the northern side of Cook Strait, a deep channel between the North and South Islands of New Zealand. Even today, this is still something of a shortcut between the Tasman Sea and Pacific Ocean for sperm whales (*Physeter macrocephalus* L.) on their travels around the region. These magnificent beasts tend to travel in groups (remember that "a number of cetacean skeletons" were found on top of the cliff). The average size of the male is approximately 52 ft (16 m) long [one of the skeletons on top of the cliff was more than 60 ft long (18 m) and so was probably a male], and mature sperm whales weigh approximately 45 tons (41 tonnes).

That is a lot of weight to be throwing around, but then iconic images of ships deposited on the top of buildings show that tsunamis have no real trouble moving heavy objects around, especially if they have a bit of air inside them. But is there a source for such a tsunami in Cook Strait? Actually there is. Cook Strait is a large submarine canyon system approximately 4000 ft (~1200 m) deep, and recent research shows evidence of numerous submarine landslides off the canyon slopes.

The picture is therefore almost complete. We have local tsunami sources, we have recent tsunamis showing us that they are quite capable of moving marine mammals around, and we have sperm whales using the area as a highway. This case seems to indicate an extremely unfortunate coincidence whereby a pod of sperm whales happened to be in the wrong place at the wrong time. Unfortunately, in the intervening century between their discovery and the rediscovery of the report, the site of the whale skeletons has been destroyed to make way for the residential subdivision.

We will probably never know definitively what put them there until the next time it happens. And it will happen. Some people will have a great view of it.

In Hot Water

Underwater landslides are at least a source of tsunamis that are moderately familiar, and we can tell numerous stories about them if pushed. They often occur off the side of volcanoes, such as the Hawaiian Islands in the middle of the Pacific Ocean and probably at some time most of the 20,000 or more atolls that are dotted around the Pacific Ocean as well. In fact, with regard to volcanoes, there are numerous ways of generating tsunamis that we tend to forget. In addition to flank collapses (an underwater landslide as part of the side of the volcano collapses), there are caldera collapses—when large volumes of magma are erupted over a short period of time, this removes any structural support for the crust on top of the now empty magma chamber. If this happens underwater and quickly, then the collapse of the crust into the empty magma chamber causes a tsunami. There are also pyroclastic flows—a dense, destructive mass of very hot ash, lava fragments, and gases ejected explosively from a volcano and typically flowing at great speed downhill into the sea. In addition, there is the really nasty explosive type called phreatomagmatic eruptions, which occur when magma and water come together suddenly—not nice, they don't mix well. For this piece, however, we use an example of something generally associated with volcanic regions but normally considered insufficiently violent to generate tsunamis—hydrothermal explosions, which are more commonly known as geyser eruptions.

In 1900, the world's biggest ever geyser, historically at least, formed in Rotorua in the middle of the North Island, New Zealand. It proceeded to erupt massive columns of boiling water up to 1,510 ft (460 m) into the air (far higher than the Empire State Building in New York). Not surprisingly, in a country desperate to attract local and foreign visitors (no problem these days due to air travel), it became an almost instant tourist attraction. It rapidly gained a name as well—*Waimangu* or "black water"—because of the sediment laden slurry it repeatedly erupted from the ground (Figure 9.1).

To put this geyser in perspective, the world's biggest active geyser today is the Steamboat Geyser in Yellowstone National Park. It is a little unpredictable but erupted in 2013, 2014, and then a few times in 2018. Steamboat, located in the Norris Geyser Basin, can shoot a column of water up to 380 ft (115 m) into the air. Although this makes it the world's biggest geyser, this is only 25% of Waimangu's efforts. Even so, the US National Park Service quite rightly publishes warnings, posts signs, and maintains boardwalks to help people view this and many other geysers in the area safely, but every year there are a few tourists who just have to get a little closer and quite often they suffer the consequences—usually severe burns, but occasionally someone dies.

In the early 20th century, such health and safety concerns were not quite as high on the Waimangu visitor experience, but even so authorities were well aware of the dangers, and although there was no barrier or protection to prevent people from going too close to the geyser, there were warning notices. Tourists flocked to see it in action, and it rather helpfully played into their

Figure 9.1 Waimangu Geyser from Frying-Pan Flat.
Source: Paul Gooding, *Picturesque New Zealand* (Houghton Mifflin, 1913, facing p. 110).

hands. It quickly established a routine of erupting approximately every 36 hours for an approximately 6-hour period. This was wonderful news for the newly created Department of Tourist and Health resorts, set up by the government specifically to promote tourism. Being close to the town of Rotorua meant that they could devise an excellent round-trip visitor experience that included the famous buried village nearby, a boat trip across Lake Tarawera, a walk to Lake Rotomahana, followed by another boat trip and a hill climb to overlook Waimangu in action. To finish the trip, visitors would take a quick stroll down the hill to catch a horse and cart back to Rotorua for the night. This was great for the tourist industry and great for Rotorua in general.

Waimangu was erupting high and regularly, getting everything out of its short life span—a life span that came to an abrupt end in 1904 following some major activity that disrupted its underground plumbing. However, Waimangu's largest eruption occurred on August 30, 1903, and in hindsight this probably marked the beginning of the end for the young geyser. On this day, there were 32 tourists on the trip—one group went straight to Waimangu, and another came the long way via the lakes and reached the geyser later in the day.

The guide for the second group was Alfred Warbrick, who had his brother Joe along to help out. Joe was famous, or about as famous as a Māori guy could be in those days. He had been a member of the first ever New Zealand rugby team that toured Australia in 1884, and he later captained the 1888–1989 New Zealand Native team that made a 107-match tour of New Zealand, Australia, and the British Isles. Among the group of tourists were Mrs. Nicholls, her two daughters Kathleen and Ruby, and David McNaughton.

According to Alfred, the geyser was very active that day, and after only 20 minutes it erupted a column of water up to 500 ft (150 m) high. From experience, Alfred knew that there was at least one more large eruption to go, and because he was looking after Mrs. Nicholls at the time, he walked her away from the geyser. Before doing so, however, he tried to warn his brother and the others. He even got Mrs. Nicholls to shout a warning, but the girls just looked around and smiled. When Arthur and Mrs. Nicholls reached a spot overlooking the geyser, they watched Joe and the others standing on the brink on the geyser. On this occasion, the eruption displaced a massive amount of boiling water well over 1,475 ft (450 m) high that was seen as far away as Rotorua some 14 miles (22 km) to the northwest. This was immediately followed by two lateral eruptions that took the form of an approximately 12-ft (3.5 m)-deep "tidal wave" that carried the bodies of Joe, Kathleen, Ruby, and David nearly 1 mile (1.6 km) in the direction of Lake Rotomahana. The hot water made it difficult to get near the victims very quickly, but when rescuers eventually did reach them, they described finding the bodies with their clothes torn to shreds by the power of the water.

This potentially unique situation may represent the only known occasion when a wave of this magnitude has been recorded from a geyser eruption. There may be some scientists out there who feel uncomfortable calling such an event a tsunami, but it is difficult to think what else to call it.

Knowing how popular geysers are with today's tourists and the never-ending search for the perfect photograph or selfie, it is perhaps important to note that one of the witnesses was informed that the victims had stayed overlooking the geyser because "one of the young ladies seemed anxious to get a snapshot."

The Shipping News

Many of us are familiar with the iconic photographs we previously mentioned of ships being stranded by tsunamis on top of buildings or far inland. In a post-2004 Indian Ocean tsunami world, we now have the *Apung*, a power-generating ship stranded 1.8 miles (3 km) inland in the city of Banda Aceh, Indonesia. This has become a well-known tourist attraction as well as a memorial to those killed by the tsunami. Ships and tsunamis have a fascinating history stretching back well into the 19th century, and we mentioned a few examples in Chapter 7. There are doubtless many earlier events yet to be discovered; here, we discuss three examples to add to our efforts in Chapter 7, but these are all slightly different.

Making a Splash

As tsunami researchers, we spend probably far too much of our time researching old written accounts of past events. This task has been made both easier and more difficult with the burgeoning of the internet. It is easier because many of these past documents are now searchable online, thus replacing much of the time spent in libraries poring over dusty tomes (although this is a rather pleasurable pastime that often turns up real gems and allows one to wander off on fascinating tangents unavailable even now in the internet age). It is more difficult because there is simply so much information online, much of it relevant and much of it irrelevant, and one often has to check it all.

This is one of the more difficult cases.

There are now numerous incredibly valuable, searchable electronic newspaper databases. You simply type in search words such as "tidal wave," "earthquake," "landslide," and so on and see what comes up. As a tsunami researcher, it somewhat grates to have to insert the search words "tidal wave," but this was the common misnomer used to describe tsunamis well into the 20th century.

Often, many results are generated. The value of such databases cannot be overstated. For example, we have managed to add at least 10 previously unrecorded tsunamis to the Hawaiian historical database alone. But these searches also call for caution. One day, as my (JG) searches of one particular database were drawing to a close, I started to add a few esoteric words such as "bomb" (big ones can cause tsunamis) and "asteroid." In the past, the latter has returned the odd gem of a find, with at least one case of a small splash (tsunami) being generated by a little asteroid plunging into the sea close to the coast of New Zealand in 1908.

This internet search involved a database of old Australian newspapers and yielded an intriguing reference to the loss of a ship and all of its crew from an asteroid strike (Figure 9.2).

The odds of such an event even happening set off alarm bells immediately, but you never know. Was it possible that this could have happened as recently as 1891 without there being more of a hue and cry than a few desultory lines in a local newspaper thousands of miles away from where the

The West Australian (Perth, WA : 1879-1954), Saturday 7 November 1891, page 5

TELEGRAMS.

£90,000 has been bequeathed to the Mildmay Mission to the Jews carrying on [illegible] London.

A REMARKABLE SHIPWRECK.

A SHIP STRUCK BY AN ASTEROID.

London, Nov. 5.

The ship *Strathblane*, bound from Honolulu, has been lost at sea. She was struck by an asteroid and sank. The captain and five of the crew went down with her.

GREAT BRITAIN AND AFGHANISTAN.

Figure 9.2 "A Remarkable Shipwreck": headline from *The West Australian*, November 7, 1891, p. 5.
Source: Trove (https://trove.nla.gov.au/newspaper/page/754781).

event is meant to have happened? There must have been more newspaper articles out there . . . and so the search began and I plunged down the rabbit hole. Sure enough, a few more brief lines of reporting were to be found in several other newspapers of the day. There were soon indications that this was a minor news story as opposed to some form of Hollywood *Deep Impact* event. A perfect example is a report in *The Mercury* (a Tasmanian newspaper) on November 7, 1891, stating that "the barque Strathblane, from Honolulu, has been wrecked at Astoria, Columbia River, and the captain and five of the crew drowned." A simple error—in western Australian the word "Astoria" was mistaken for "asteroid," and in modern parlance, Fake News was created! If there is any useful message from this story, it is undoubtedly that one should never believe everything that is written in the newspapers. It didn't change the world, but it does at least indicate that newspapers have occasionally been getting things wrong for a long time . . . and will doubtless continue to do so.

The Downside of Cruise Ships

There is nothing quite like taking a cruise. There are gorgeous sites that in many cases are simply far more accessible by ship than by plane, train, or automobile. Enjoying the views while sitting back with a gin and tonic and tucking in to yet another glorious meal without having to even search for a restaurant is also something of a bonus for many.

In the northeastern part of the Pacific, just across the border from Canada, sits the town of Skagway in what is known as the panhandle of Alaska. Historically, this is an interesting patch of land close to the border between Alaska and British Columbia. It was associated with the Alaska boundary dispute, in which the United States and Britain (which was handling Canada's foreign policy at the time) claimed different borderlines around the Alaskan panhandle. This dispute had been ongoing since 1821 and was originally between Russia and Britain. The United States took it on following the Alaska Purchase in 1867, when it bought the territory from Russia. Arbitration in 1903 saw the United States winning out, Canada lost sea access to the Klondike goldfields, and the British government effectively alienated itself from Canada for betraying it in favor of preserving healthy Anglo-American relations—politics, gotta love it.

Skagway, the original starting off point for the Klondike gold rush, sits at the head of a beautiful long fjord to the northeast of Glacier Bay National Park. In 1994, work was being done to renovate the railroad dock and ultimately convert it into an area for cruise ships so that the town and region could take advantage of the growing tourist industry. At the time, the dock was piled high with rock and soil being used to fill in a space between the dock and railway embankment. Also on the dock were all of the bulldozers and pile-driving equipment necessary to complete the task. The combined weight was enormous, approximately 5 million pounds (>2 million kg). Just the week before the events detailed here unfolded, one of the workmen had noticed a crack in the ground near the shoreline. Concerned about safety, this wise man simply quit his job and walked off the site. As is common with such stories of human folly, more warning signs went unheeded (see Chapter 6 regarding the Vajont Dam). The following week, on November 2, there was even more land movement, which caused some of the braces being used to shore up the ground at the construction site to break, but work went on as normal.

Two days later at 7:10 p.m. on November 4, there was an exceptionally low tide, effectively raising the whole site even farther above sea level than normal. Paul Wallen of Homer, Alaska, and his brother were working inside a steel caisson on the site when suddenly the steel sheet pilings snapped in two and debris began raining down on top of them. Paul's brother managed to scramble to safety, but Paul was trapped inside. The sediment supporting the heavy dock gave way and slid into deeper water. Then the old dock, with everything on it, including the bulldozers, was carried away. Eight hundred feet (240 m) of the railroad dock suddenly vanished. The underwater landslide was some 600 ft (180 m) wide, at least 50 ft (15 m) deep, and composed of between 1 and 3 million cubic yards (760,000 to 2.3 million m^3) of earth. The harbor had been instantly deepened from 40 ft (12 m) to more than 90 ft (27 m) in places. Enough water was displaced to generate 20-ft (6-m) tsunami waves that heavily damaged Skagway's ferry terminal and snapped a fuel line, sending gallons of fuel oil into the water.

Farther up the inlet lay the Skagway Small Boat Harbor. On one of the boats in the harbor, two men turned to see the tsunami heading toward them and began to dash up the floating dock to the shore. But once the tsunami struck, boats began smashing into each other, and the dock was twisted and buckled. The men could now only crawl toward the shore, but ultimately they made it to safety, injured but alive. Paul Wallen, trapped inside the steel caisson back at the railway dock, would not be so lucky. His body would not be found for nearly 3 months.

Unlike the 1964 earthquake (see Chapter 8), no earthquake had triggered the slide—this was a nonseismic tsunami. Although such events are relatively rare, they are a serious hazard because there is no natural warning—no ground shaking to alert people to flee. The tsunami had no Pacific-wide consequences, yet it still proved to be a killer, and it remains the most recent deadly tsunami to strike Alaska.

This may get repetitive but—the next one can happen at any time.

Halifax Harbor—Blown to Hell

Unfortunately, human folly knows no limits. World War I was raging in Europe and the port of Halifax, Nova Scotia, was the closest North American port to the war zone. Being closest meant the shortest ocean crossing, and that meant less chance of Allied ships being sunk by German U-boats. As a result, the port served as an important stopover for ships carrying supplies and troops from North America to Europe and for returning ships on their way to reload. It was an efficient system that worked well until the fateful morning of December 6, 1917.

En route from Holland to the United States, the former Norwegian whaling supply ship, SS *Imo*, was making a brief stopover in Halifax. She had recently been chartered as a "Belgium Relief" vessel and was rushing to New York to load up emergency food rations to carry to thousands of starving civilians in northern France and Belgium, whose food had been confiscated by the German army. Rather ironically, the *Imo* was built in the same shipyard as the ill-fated RMS *Titanic* that had sunk less than 5 years before, and the Fairview Cemetery in Halifax, a mere quarter of a mile from the sea, was the final resting place for more than 100 of the *Titanic*'s victims.

Also in port that day was the French vessel, SS *Mont-Blanc*, preparing to transport a very different cargo for the war effort. The *Mont-Blanc* was loaded with explosives including gun cotton, picric acid, and 5,000 cases of TNT, equivalent to 2,900 tons (~2,600 tonnes) of TNT. A former merchant ship, she was suffering from years of use with minimal maintenance, and with the coming of war had been quickly converted into a navy munitions vessel with a couple of guns mounted on her deck, more for show than as effective deterrent against U-boats. She had also been painted a camouflage gray, making it more difficult for the ship to be seen—a useful deterrent against U-boats and other, presumably enemy, ships.

Just before 9.00 a.m., the *Imo* gave a single horn blast as she hurried toward the harbor entrance heading for New York through what the local indigenous *Mi'kmaw* people accurately called the *Kepe'kek*, "the Narrows." The crew on the *Mont-Blanc* hadn't heard the single horn blast and only saw the *Imo* steaming rapidly in their direction when she was less than half a mile away. Normally, ships pass each other on their port (left) side, but at that exact moment there was more room to pass on the starboard (right) side. The crew of the *Mont-Blanc* looked on in awe and then horror as the *Imo* followed normal protocol and tried to pass on the port side along what was quite literally a collision course.

Before any corrective action could be taken, the *Imo* smashed into the starboard hull of the *Mont-Blanc*. Drums of highly combustible fuel stored on the upper deck of the *Mont-Blanc* (away from the TNT) burst into flames. Her captain and crew, aware of their extremely dangerous cargo, quickly jumped into their lifeboats to escape what they feared would happen next. They raced to land, jumped ashore, and literally ran for their lives.

Meanwhile, lacking a crew, the burning *Mont-Blanc* drifted toward shore, running aground at 9:04 a.m. As soon as she hit the shore, her cargo violently exploded. The incredible heat of the explosion instantly vaporized the water around and under the ship, leaving a huge vacuum. When the water rushed in to fill this void, it created a tsunami. Near the site of the explosion, survivors reported seeing three "tidal waves," one "more than twenty feet above the level of the harbour" and then "it emptied Halifax Harbour right out . . . just a little stream of water down the middle on the bottom." In the harbor, but away from the Narrows, the tsunami run-up was probably in the range of 6–12 ft (1.8–3.6 m). But within the Narrows, the water reached much greater heights, achieving its maximum run-up of 60 ft (18 m) at Dartmouth just opposite the explosion site.

Phillip Mitchell was on the Dartmouth shore at the time of the explosion. He ran to an electricity pole and, seeing "a huge wave," wrapped his arms about the post and held on. The water surged over his head and soaked a boxcar on the railroad track above him, but he managed to hold on and survive. The wave then receded, only to be followed by a second smaller wave that fortunately "only reached his arm pits."

Tufts Cove, adjacent to Dartmouth, also felt the full impact of the tsunami. Several houses were destroyed and 9 people were killed by the waves. Around the harbor, the explosion and tsunami instantly obliterated roads and railroads, killing 61 train crew members and destroying more than 500 train cars. It cut telephone and telegraph lines and destroyed submarine cables that passed through the Narrows, effectively breaking communications between Nova Scotia and continental America, as well as the rest of the world (submarine cables are discussed in Chapter 16). Approximately 10 minutes after the explosion, "black rain" fell from the sky as an oily soot of unconsumed carbon from the explosives came down to earth. It blackened faces, penetrated clothing to the skin, and coated the ruins of the houses. The final casualty list totaled 2,000 dead and 9,000 injured (Figure 9.3).

This was the largest man-made explosion at that time in history and would be studied by the super-secret US Manhattan Project to estimate the potential blast radius of atomic bombs. The Halifax explosion would remain the greatest man-made explosion in history until the United States dropped the atomic bomb on Hiroshima, Japan, on August 6, 1945.

Amazingly, the *Imo* would be refloated, repaired, and become a whale oil tanker. But not for long, because she ended up on the rocks off the Falkland Islands in 1921. Her captain when she was in Halifax in 1917 was not in command at the time of her demise because he was one of the 2,000 victims who died in the explosion.

To this day, the Halifax explosion remains the deadliest disaster in Canadian history, and every year on December 6, a memorial ceremony is held in Halifax to honor the victims of the explosion and tsunami.

Figure 9.3 A view across the devastation of Halifax 2 days after the December 6, 1917, explosion, looking toward the Dartmouth side of the harbor. The SS *Imo* is visible aground on the far side of the harbor.

Source: https://commons.wikimedia.org/wiki/File:Halifax_Explosion_-_harbour_view_-_restored.jpg.

Alcohol and Tsunamis Don't Mix

We know that alcohol and many things don't mix. As the old adage goes, "There's many a slip 'twixt the cup and the lip."

One often wonders why we are bludgeoned with so many health and safety warnings these days. Compliance has become a dirty word, especially among those of us who have to comply with what often seem to be extremely petty demands in this day and age. The following human-induced tsunami stories do at least show the value of some of these controls and the fact that we have indeed learned from our mistakes.

The Great Gorbals Whisky Flood of 1906

The Gorbals is an area on the south bank of the River Clyde in Glasgow, Scotland. It was not the most salubrious place, and by the late 19th century it was a densely populated, densely industrialized part of the city. Due to a declining industry and overpopulation in the area, it became a center of poverty.

As the jobs declined, Gorbals became one of Glasgow's more dangerous slum areas associated with drunkenness and crime. Indicative of this decline was the short-lived railway station, opened in 1877 and closed in 1928—Gorbals was not a wonderful place to be living at the turn of the 20th century.

The Loch Katrine or Adelphi Distillery was built in the area in 1826, and at the time of the disaster it was one of the largest distilleries in Scotland, with an annual output of more than 500,000 gallons (2.3 million liters) of whisky. The process of whisky manufacturing or creation is somewhat convoluted, but here we are interested in only the early stages. Malt is first "mashed," which generates a lot of heat. The result of mashing the malt—the basic ingredient of whisky—is a liquid known as wort, which contains the sugars that are fermented by yeast to produce alcohol. Because yeast can't stand hot temperatures, the wort is cooled to a warm 20°C before it is mixed in washback vats.

Early in the morning of November 21, 1906, one of the distillery's massive washback vats stored on the top floor (a wise location?) of the distillery collapsed. The knock-on effect was the collapse of two more adjacent vats, releasing a total of approximately 150,000 gallons (680,000 liters) of hot whisky—approximately 7–10 proof at the time. It flowed down through the building into the basement and onto the street outside.

This veritable tsunami of whisky engulfed a number of carts and their drivers, throwing them across the street, where they struggled to escape the waist-deep wave of alcohol. As this alcoholic brew had cascaded through the distillery, it had picked up a vast quantity of draff from the basement. This draff, a husk residue of the fermented grain used as a food for cattle, helped create a fluid akin to the consistency of liquid glue. Although the police were quickly on the scene and managed to rescue most people, some were in a poor state. The force of the hot whisky–liquid glue mix tore most of the clothing off one man. It struck a bakehouse near the distillery, throwing one man against the wall, and the staircase collapsed, trapping four men upstairs who had to jump out of windows to escape. The wave rampaged through 64 Muirhead Street, washing the elderly Mary Ann Doran off her feet, and it was only after several attempts that she was able to escape through the front door. Although these were some of the many lucky escapes—something we often hear about in the aftermath of a tsunami—there was one fatality. James Ballantyne, a farm servant, suffered severe internal injuries and died soon after he was admitted to the infirmary.

Although some might argue whether this could be considered a tsunami or not, it was very much akin to a small-scale version of the gravity-enhanced dam-burst flood of the Vajont Dam, Italy, in 1963 (see Chapter 6).

The Loch Katrine Distillery closed the following year in 1907.

Boston Molasses Flood of 1919

January 15, 1919, was a very mild day for Boston in mid-winter. During the previous 3 days, temperatures had risen from approximately 2°F to 43°F (–16°C to 6°C). In the area around Copps Hill were the freight sheds of the Boston and Worcester and Eastern Massachusetts Railways, and looming large over these was a huge 2.5-million-gallon (9.7-million-liter) tank containing molasses adjacent to the relatively new elevated railway—the "El." The tank had been built 4 years earlier by the Purity Distilling Company but was now owned by the United States Industrial Alcohol Company. This was a massive construction comprising curved steel sides held together with rivets and huge bottom plates set into a concrete base. Although the origins of New England's and hence Boston's demand for molasses were born out of the old Colonial triangle trade of slaves from Africa to the West Indies, molasses from the West Indies to New England, and rum (made from the molasses) back across the Atlantic to purchase a cargo of slaves, this format was long gone by 1919. However, rum was still made in Boston—and incidentally baked beans as well—and the molasses needed for the task still came from the South.

The United States was about to enter one of its more interesting social experiments with Prohibition about to be ratified with a vote the next day. It is entirely conceivable that the Purity Distilling Company could read the writing on the wall and so by selling out in 1917 to the United States Industrial Alcohol Company it ensured that the huge molasses tank, 50 ft (15 m) tall and some 90 ft (27 m) in diameter, could continue to supply the raw ingredients for a legal alcohol trade—to industry. Indeed, business was good, and just a few days before, a ship from Puerto Rico had bulked up the contents of the tank to approximately 2.3 million gallons (8.7 million liters).

It was a glorious day for the middle of winter, and people were lounging around in their shirtsleeves. The general cheer of a nice warm day was bringing smiles to all around, and lunchtime was fast approaching.

At approximately 12:30 p.m., a low, deep rumble reverberated through the freight yard. The ground seemed to heave under people's feet, and the giant molasses tank tore apart. It appeared to rise up and split, with the steel rivets popping like machine-gun fire, releasing a sticky brown geyser up into the sky under a pressure of 2 tons per square foot (191.52 kPa). This caused a tsunami of choking molasses, 15–25 ft (4–7 m) high, to flood the immediate neighborhood of downtown Boston. It generated a hissing, sucking sound as it splashed in a curving arc, crushing everything and everybody in its path. Men, women, children, and animals were caught, hurled into the air, or dashed against freight cars only to fall back and sink from sight in the slowly moving mass.

A steel section of the tank was hurled across the street, taking out one of the vertical supports for the El. As the track ahead sagged into the onrushing molasses, an approaching train screeched to a stop, just avoiding an even more devastating disaster. The wave moved at approximately 35 miles (55 km) per hour. Many of the lunchtime crowd of laborers were crushed or drowned, and schoolchildren on their way home from morning classes were also caught by the wave. Some survived. One child picked up by the wave was carried almost like a surfer on its crest only to be rolled off later on. His throat was so clogged with molasses that he was thought dead until he opened his eyes. Later, a wagon driver was found solidified in death like a figure from Pompeii.

It was all over in a matter of minutes, with the molasses covering several blocks of downtown Boston with a 2- to 3-ft (up to 1 m) "tsunami" deposit. Although rescue equipment was quick to arrive on the scene, rescuers had to place ladders over the wreckage and crawl out on them to pull the dead and dying from the molasses-drenched debris. It took time. It was too sticky to walk through, and ultimately this cloying gooey mess was cleaned up by hosing the area with salt water and then covering the streets with sand.

The final toll was 21 dead and 150 injured. After a long, protracted series of lawsuits and litigation—this is the United States after all—the courts eventually found that the tank had ruptured because it was incapable of withstanding the pressure of the load it held. Simply stated, inspections were less than rigorous and the company was held responsible.

In an ironic twist, during the night of January 16 as people continued to tackle the disaster and the incredible mess caused by one of the ingredients for the making of rum, they were interrupted by the sudden ringing of church bells throughout downtown Boston. Prohibition was now law, and churches were celebrating.

Hmmmm . . . there is probably a moral in this story somewhere.

The Tsunami Bomb

Project Seal doesn't sound like a particular auspicious name, but it was important enough to have been a major secret that didn't really see the light of day until quite recently in 1999. Only 160 copies of the original report related to this project were ever made, and there was a bit of a crisis at one point as the authorities tried to determine where they all were—had the secret leaked out?

Why all the fuss?

During World War II as the US Navy struggled through the Pacific in a brutal conflict with the Japanese, it needed to create better anchorages for its vessels. US Navy officer E. A. Gibson noticed that quite often when the Navy used a large quantity of explosives to blow up the coral reef around islands to establish shipping lanes, this would often create large waves.

Based on this observation, Project Seal was established in 1944 to determine whether a tsunami bomb could be developed. Could it make really big waves? Tests were performed within the Pacific theater of war, but notably away from US shores. So a friendly (or gullible?) ally was called in, and the concept was jointly developed by both the US and New Zealand military.

The project was led by Thomas Leech, who had been appointed Professor of Engineering at Auckland University College in New Zealand in 1940. He was seconded into the army in order to carry out this top-secret work. To put this in context, Project Seal was considered so significant that it rivaled the development of the atomic bomb. It was anticipated that such a bomb would have potential to cause massive damage to coastal cities and their defenses. Although no specific targets were ever mentioned, it has been suggested that the bomb might have been used prior to invasion of the islands of Japan.

Leech carried out underwater explosions off the coast of Auckland near Whangaparaoa Peninsula and also near New Caledonia. Roughly translated, Whangaparaoa is Māori for "Bay of Whales," and although today pods of orca and dolphins are regularly seen in the waters off the peninsula, there is no record of the effects these explosions had on them at the time. This was wartime, and such considerations were not taken into account.

Although Leech carried out approximately 3,700 test explosions off Whangaparaoa alone during a period of 7 months, they were all fairly small-scale. There was no desire to inundate home turf with massive waves. Ultimately, he showed that a single explosion would not produce a tsunami but that a line of 4.4 million pounds (2 million kg) of explosives approximately 5 miles (8 km) off the coast could create a destructive wave. Although this research was completed in January 1945, work pressure on Leech meant that the university did not release him to complete his analytical work until 1950. He finished his final report in December 1950, too late for World War II, and definitely too late for it to be considered particularly important anymore.

A recent examination of the report's details by a colleague at the University of Waikato in New Zealand indicated that if this were tried today, it may well be capable of generating waves up to 100 ft (30 m) high! Although it should be noted that the huge technical difficulties associated with laying the explosives in the correct water depths, among a host of other variables, would make it an impossible engineering feat—let alone trying to do it covertly in wartime conditions.

It is worth noting as an afterword here that the US Armed Forces learned from this work, and in 1971 the US Army Engineer Nuclear Cratering Project created a harbor at Kawaihae on the Big Island of Hawaii using approximately 12 × 10-ton (9-tonne) charges of ammonium nitrate—this was code-named Project Tugboat. The group created a channel varying in width from 150 to 260 ft (45–80 m) and a 400-ft (120-m) square berthing basin with a minimum water depth of 12 ft (3.5 m). The group also generated a tsunami up to approximately 4 ft (1.2 m) high. No research has ever been carried out on the deposit created by this tsunami, but not surprisingly, it is known that a huge amount of coral debris was generated by the explosions, so it must be there somewhere.

Off the End of the Runway

It seems fitting to end this chapter in a slightly exotic location since we have wandered the seedy streets of early 20th-century Glasgow and been blown up in Halifax. Well, at least we can pretend for the briefest of moments that it is exotic.

When people hear the name "French Riviera," they think of beaches frequented by movie stars, although those times are long past and today they are covered with an assorted bunch of tourists. Either way, it would not have been a good place to go for a stroll on October 16, 1979. On that fateful day, two waves inundated a 20-mile section of the Mediterranean coast of France, stretching from the Italian border to the ancient town of Antibes, where waves approximately 10 ft (3 m) high came crashing ashore, washing nearly 500 ft (150 m) inland. This completely unexpected tsunami caused considerable damage and resulted in 10 deaths.

To say that this was completely unexpected is no exaggeration, and to this day there is still some uncertainty in the scientific community regarding exactly how the tsunami was produced. But we have a key suspect—a landslide of ground fill material that was being used to extend the airport runway at Nice. This landslide fell into the sea and generated a tsunami; however, there were also two large submarine landslides (perhaps caused as a result of the one off the end of the airport runway) on the continental slope just offshore, which continued as turbidity currents under the sea, cutting submarine cables at distances of 50 and 60 miles offshore from Nice. The tsunami may have been produced by the airport landslide; the airport landslide may have triggered the larger, deeper submarine landslides, which could have generated the tsunami; or try to come up with a permutation or combination of these and you may be correct.

Summary

As tsunami researchers, we continue to try to understand all of the possible ways in which tsunamis can be generated, and yet there is an almost endless litany of unexpected events, such as those discussed in this chapter. We have now traced unexpected events from whales on cliffs to whisky in Scotland and, finally, the French Riviera. Chapter 10 takes us to an entirely new level of unexpected, and it's a big one.

10 Megasharknado

> *Carcharocles megalodon* ("Megalodon") is the largest shark that ever lived. It has been suggested that this species was a cosmopolitan apex predator that fed on marine mammals from the middle Miocene to the Pliocene (15.9–2.6 Ma). Prevailing theory suggests that the extinction of apex predators affects ecosystem dynamics. Accordingly, knowing the time of extinction of *C. megalodon* is a fundamental step towards understanding the effects of such an event in ancient communities.
>
> **—Pimiento and Clements (2014)**

Disturbing an Ecosystem

This story relates to one of those strange coincidences in science that occur when you are doing research on one thing and some snippet of information wanders across your desk.[1] My partner is doubtless getting fed up with hearing me say something like "Oh, that's interesting," a statement that is invariably followed by an hour or so of frenzied reading that is often a disappointing dead end, but on occasion one can find an enormous rabbit hole full of exciting new information. So, here we go, into the rabbit hole.

At up to 98 ft (30 m) long, the megalodon or "megashark" was big and fast, a perfect subject for a scary Hollywood movie (Figure 10.1). It preyed upon many cetacean species, such as dolphins and small whales, so not surprisingly the species that we think of today as being particularly terrifying, such as the great white shark [the largest is ~20 ft (6 m) in length], avoided areas where the megalodon hunted. The megalodon went extinct approximately 2.6 million years ago (although there is some debate about that)—this is perhaps better known as the start of the Ice Ages when global climate cooled and widespread glaciation started in the Northern Hemisphere. Recent research has suggested that the megashark might have gone extinct slightly earlier, but this

[1] We thank Professor Mike Archer, UNSW Sydney, Australia, for suggesting the title of this chapter and for endless discussions on the topic.

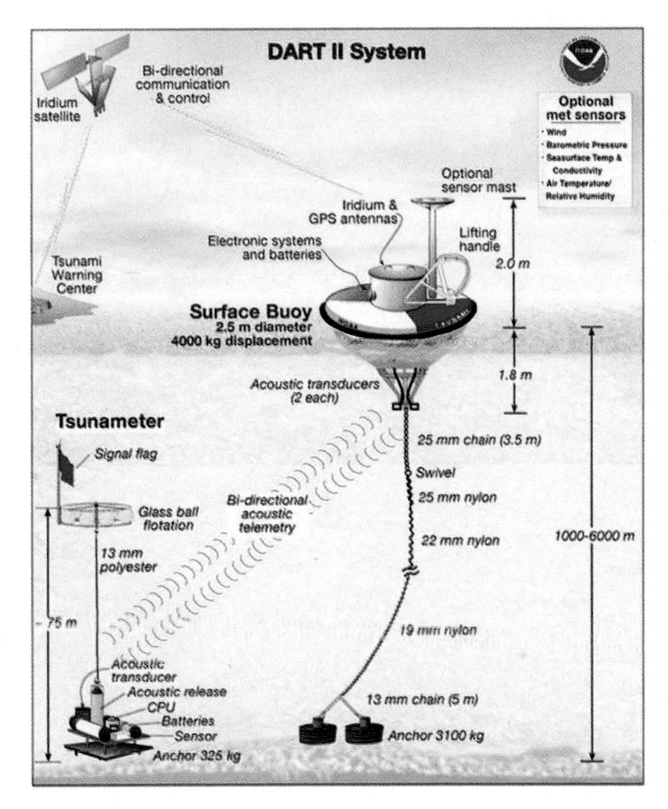

Figure I.1 Diagram of the Deep-ocean Assessment and Reporting of Tsunamis (DART) II buoy.

Source: National Oceanic and Atmospheric Administration, Pacific Marine Environmental Laboratory.

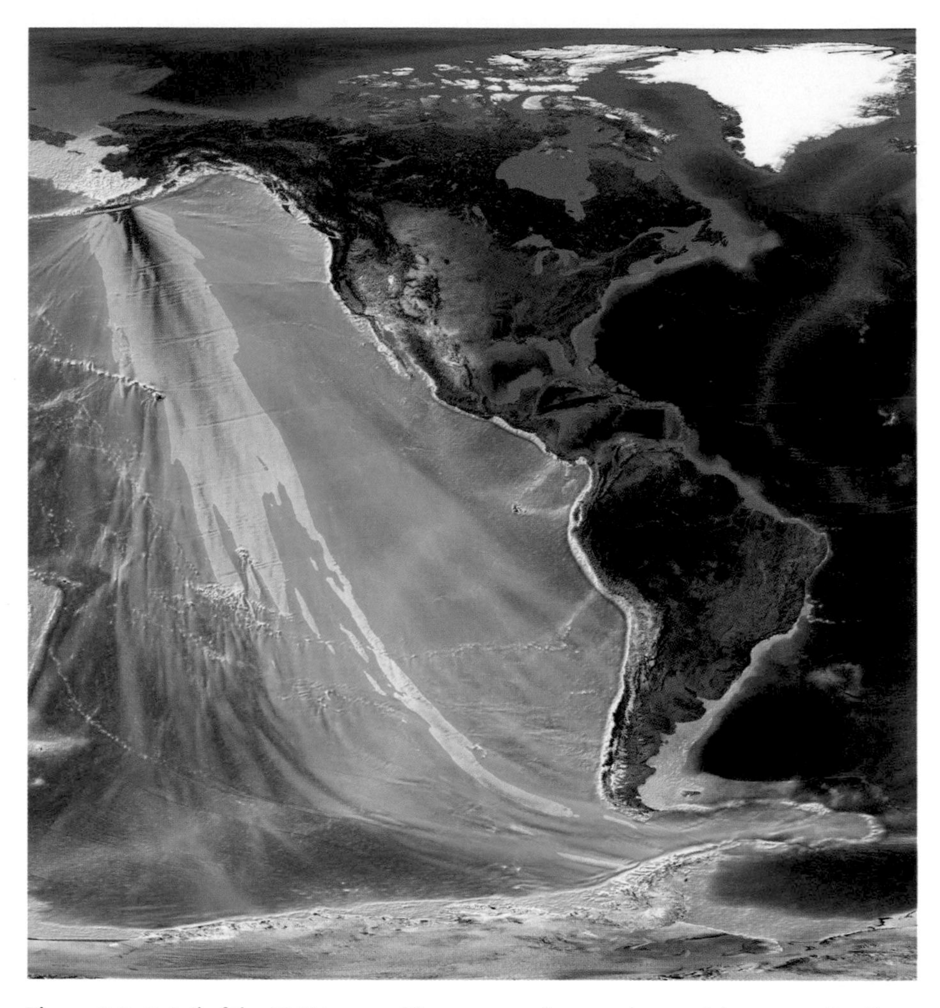

Figure 2.2 Detail of the 1946 tsunami "energy map" computer model representing the maximum rise in sea-level on the open ocean caused by the tsunami. It was far more severe in the middle of the "beam" of energy than on its sides – larger wave struck coastlines more directly in the "beam".
Credit: NOAA. https://sos.noaa.gov/datasets/tsunami-historical-series-aleutian-islands-1946/

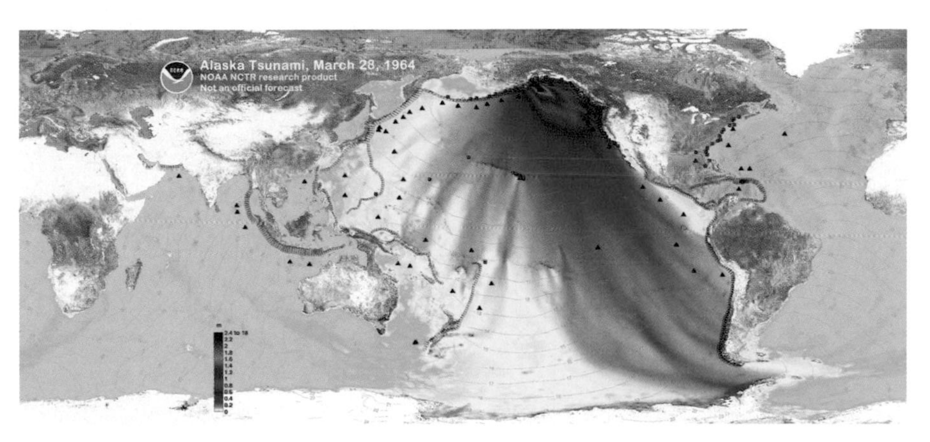

Figure 8.1 Global maximum tsunami wave amplitude for the 1964 Alaska tsunami. The black triangles show where DART (Deep-ocean Assessment and Reporting of Tsunamis) buoys have been deployed during the past 20 years for the early detection, measurement, and real-time reporting of tsunamis in the open ocean. The snake-like lines are major subduction zones.

Source: National Oceanic and Atmospheric Administration/Pacific Marine Environmental Laboratory/Center for Tsunami Research.

Figure 11.1 East Beach, Palaikastro, Crete. It may not look like much, but the entire upper part of this small cliff is the Santorini tsunami deposit; the white stick to the left of the photo is 1.0 m long.

Source: Professor Hendrik Bruins, July 24, 2006 (used for the annotated Figure 10 in the publication by Bruins et al., 2008).

Figure 13.1 (a) Dogger Bank—the "upland" area of Doggerland. (b) Image of the area known as Doggerland, which connected the British Isles and the European continent.
Sources: (a) NASA image NorthSea1.jpg, brightened and with the outlines of the Dogger Bank added (the latter taken from de:Bild:Doggerbank_Plan.jpg). (b) Image created by Max Naylor (retrieved from https://commons.wikimedia.org/w/index.php?title=File:Doggerland.svg&oldid=329399527).

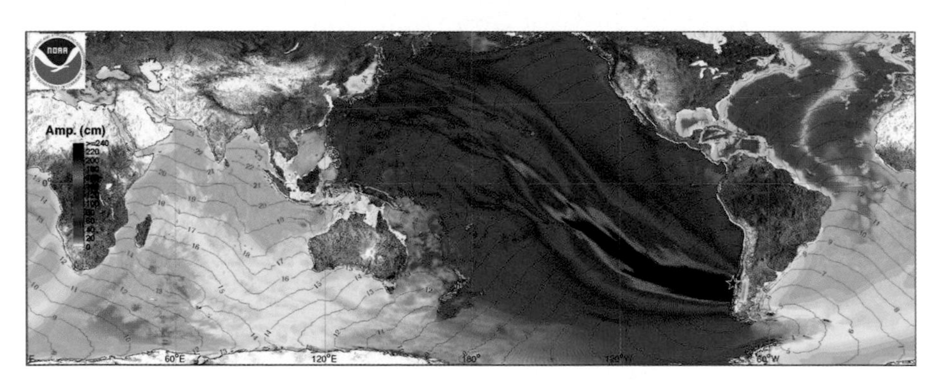

Figure 14.1 Detail of the National Oceanic and Atmospheric Administration's model of the wave amplitudes for the 1960 tsunami using historical data.
Source: NOAA Center for Tsunami Research.

Figure 15.1 Resort staff at Golden Buddha Beach, mesmerized by the approaching wave seen in the background.
Source: Jason Beloso.

Figure 15.2 Tsunami striking Golden Buddha Beach as seen from lookout above the beach.
Source: Kimina Lyall.

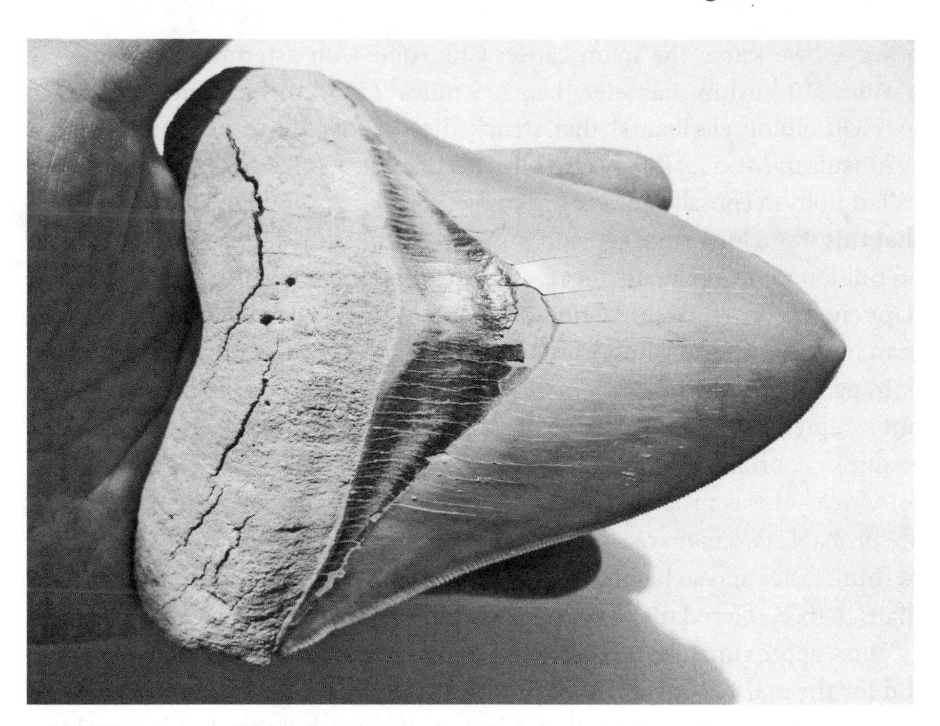

Figure 10.1 Megalodon (*Carcharocles megalodon*) tooth [5.4 inches (13.7 cm) long and 4.4 inches (10.2 cm) wide] excavated from Lee Creek Mine, Aurora, North Carolina.
Source: Wikipedia Commons (https://commons.wikimedia.org/wiki/File:Carcharocles_megalodon_tooth.JPG; attribution: Tomleetaiwan).

is largely based on poor dating for some of the fossil remains, so the jury is still probably out on this one.

In addition to the megashark, there is much evidence for other things going extinct approximately 2.6 million years ago, and this is where all the odd snippets of information started to look interesting. There was a huge extinction event in the sea that affected mammals, seabirds, turtles, sharks, shellfish, and a wide range of microorganisms—36% of the existing genera simply didn't make it beyond this point. Whereas some seem to have gone extinct immediately, others slowly died out as the waters got colder; in effect, they were simply frozen out. An interesting consequence of the death of the megashark, however, was that the blue whale and other marine species grew in size because they were no longer being eaten by this apex predator.

It is important to mention that this mass extinction pales in significance to similar events, such as the one that occurred at the end of the Cretaceous approximately 65 to 66 million years ago. This saw the end of the dinosaurs and more than 60% of all of the species in existence at that time. In that case, however, we know the main cause: Chicxulub—an asteroid approximately 6 miles (10 km) in diameter [i.e., 125 miles^3 (523 km^3) around; equivalent to ~550 billion elephants] that struck the shallow ocean near the Yucatan Peninsula in Mexico. It may also have been accompanied by a smaller brother called Boltysh (possibly 1 or 2 km in diameter) that hit the Ukraine. It is true that this was a long time ago and it is difficult to truly understand what 65 to 66 million years ago really means, but the dinosaurs died out then, and what happened approximately 2.6 million years ago, much closer to the present day, is an indication of how important it is to study these extinction events.

In geological terms, 2.6 million years ago is maybe a few blinks of the eye, but it represents a time period that starts to impinge upon us as humans. For example, approximately 2 million years ago, our early ancestors migrated out of Africa—it was probably *Homo erectus.* Why did they migrate then? There are probably two main reasons: (1) They had evolved to the point of being able to think faster and so be more adaptable to change; and (2) such a change, climate change, offered the opportunity for this to happen.

Thus, approximately 2.6 million years ago there was a mass extinction that did for the marine world what Chicxulub did for the dinosaurs. Because this was notably a marine extinction event, it was much better hidden geologically than if it had happened on land. Interestingly, however, not long after it occurred our ancestors started to move out of Africa, eventually populating the entire planet. Therefore, the major question is, What happened? Surely because it was such a relatively short (geological) time ago, we must know much more about this than we do about Chicxulub and the end of the dinosaurs. Amazingly, we don't, or rather perhaps we do. The problem is that because we know much more about geological and ecological changes that were occurring at this time, there is a lot of "noise" in the data. In a sense, we can't see the forest for the trees.

Several ideas have been put forward about what happened approximately 2.6 million years ago. Some scientists have noted that coastal areas were moderately hostile environments for biota, probably because the sea level fell rapidly as a lot of water got tied up in ice sheets and/or carbon dioxide concentrations tumbled. This is very interesting but simply goes to show that the climate cooled quickly; however these are effects—something had to make them happen . . . and quickly. In addressing this speed of change, some geologists have noted the geologically rapid uplifting of the Rocky Mountains and the Himalayas. Such changes would have fundamentally changed weather patterns. In addition, the Panama Isthmus formed completely, joining North and South America together (a canal later undid much of that hard work), and this also changed ocean circulation patterns.

These are prominent arguments in the discussions surrounding events that occurred approximately 2.6 million years ago, but there is one significant problem that may simply represent how geologists view the world. These rapid geological changes that definitely occurred around that time are just that—geologically rapid. These do not explain the rapid mass extinctions in the marine realm. What is needed is a change that has an almost instantaneous effect on the marine ecosystem, and those that are proposed are not fast enough. It is a bit like comparing the movement of a snail with that of a space rocket, although there is a vigorous argument that these gradual geological changes eventually led to a sudden ecological switch being flipped that caused these extinctions. In other words, eventually the environment passed beyond a threshold where megasharks and many other species could exist and so they all died out at approximately the same time. It is indeed plausible that these geological explanations provide possible solutions, or part of the solution, for the mass extinctions. Equally, however, these all seem to be somewhat convoluted geological explanations for what was a sudden change that simply cannot be explained with the data available to us at the moment. Or can it?

Enter the Tea Baron (El Tanin . . . A Sad Pun)

A few voices in the academic literature have proposed an alternative hypothesis, one that is gaining some traction. The basic question has been, What about an asteroid? Enter Eltanin, or as a colleague more informally calls it, the Tea Baron (tannins give teas their bitter taste, so El Tanin sounds like a good name for a drug lord like a tea baron . . . or not).

Eltanin struck the Earth approximately 2.5 million years ago ("approximately 2.5" and "approximately 2.6" are used here to smooth over uncertainties in precise dates—essentially they are the same date), at precisely the time that geologists agree that the Quaternary began. This is when ice sheets started to form and the megasharks and many other marine species died out. Is this a coincidence?

First, it is important to review what we think we know about this event, or perhaps the best way of saying it is, let's have a look at the tantalizing clues we have found so far. As it stands, the Eltanin asteroid is currently the only known deep-ocean impact in the world, although there may be a few other candidates. "Deep ocean" means that it struck in water depths of approximately 2.5–3.0 miles (4–5 km) [probably at ~30 miles/second (~50 km/second)]. Compare this with a depth of approximately 110–165 yards (100–200 m) for the Chicxulub event and you get a sense of how massive a depth of water is involved. Research on the Chicxulub is largely responsible for the coining of the term "mega-tsunami," which recognizes the extreme wave heights that can be generated when an asteroid hits the ocean (Figure 10.2). Tsunami wave heights where Chicxulub struck were approximately 110 yards (100 m) or more. The tsunami couldn't have been much bigger because it only had that depth of water to move.

At this point, it is important to make a mental note: This means that if an asteroid strikes in water of 4 or 5 km depth, then there is the potential to generate a 4- or 5-km-high wave. Let us not forget that the Earth is approximately 70% ocean. Therefore, asteroids are far more likely to hit the ocean than the land, but the problem we have is finding evidence for them because all the evidence is underwater.

Eltanin struck in the Southern Ocean approximately 930 miles (1,500 km) south–southwest of Chile—the middle of nowhere. There is no crater on the seafloor. This means that because the ocean is 2.5 or 3.1 miles (4 or 5) km deep at that location, the asteroid can't have been larger than that, so it was therefore much smaller than Chicxulub. In fact, the Eltanin asteroid is considered to have been 0.6–2.5 miles (1–4 km) in diameter. That's still pretty big—at its largest, most humans would take a good 40–45 minutes to walk that distance. At the top end, this makes its volume approximately 8 miles^3 or

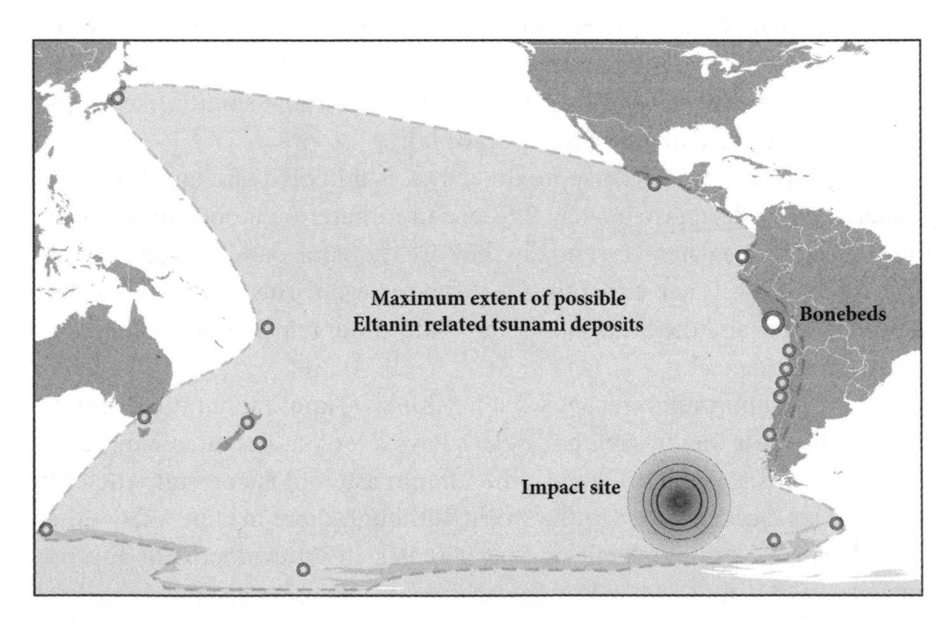

Figure 10.2 Possible extent of the Eltanin mega-tsunami
Source: After Goff et al. (2012).

33 km^3—"only" approximately 35 billion elephants in the old money we used previously.

If there was no crater, however, how do we know the asteroid even struck? This was pure luck. It was one of those fortuitous events that occurs from time to time to give scientists "eureka" or "that's interesting" moments. The USNS *Eltanin* (after which the asteroid is named) was a cargo ship with an ice-breaking hull. The story of the *Eltanin* is worthy of an entire book, but it only receives a brief moment of fame here. It was acquired by the US Navy in 1957 and then became an oceanographic research ship in 1962. It was used around Antarctic waters for a decade or so with a brief stint with the Argentinian Navy before being scrapped in 1979. This was before scientists had even discovered what she had pulled up from the seafloor.

Evidence for the Eltanin asteroid was finally discovered in 1981 when scientists noticed odd geochemical and sediment signatures in cores more than 310 miles (500 km) apart that had been extracted from the seafloor by scientists onboard the ship in the mid-1960s. A later research trip to the same areas in 1995 put more flesh on the bones of these early findings. The fact that there was no crater meant that scientists could basically state, " 'OK, this is as big as it could have been." The irresistible force of a large rock up to 4 km in diameter hitting the almost immovable object of the deep ocean resulted in the asteroid shattering/melting/vaporizing/pulverizing on impact, while it also created a huge, temporary "crater" in the ocean. In essence, the asteroid and the ocean canceled each other out.

What happens when a huge hole is made in a vast expanse of ocean? People tend to remember splashing around in a bath when they were young, plunging a large object (a plastic megashark or elephant perhaps) into the bath (Pacific Ocean), resulting in lots of water splashing up and parents being annoyed as the bathroom floor was flooded as the 'tsunami' that was generated overtopped the edge of the bath. In the Tea Baron's case, the tsunami could have been up to 2.5 miles (4 km) high at that point of impact . . . gulp.

Piling on the Pressure

Let us do some back of the envelope calculations and try to establish Eltanin's place in history. First, an asteroid impact in the range of 0.6–6.0 miles (1–10 km) in diameter may raise enough dust into the stratosphere to cause a global catastrophe, leading to mass starvation, disease, and general disruption of social order—and that's just the effect on humans. We'll return to that later.

At 2.5 miles (4 km) in diameter, the impact would have delivered an estimated energy equivalent to approximately 650 million Hiroshima bombs [~6,400 exajoules (EJ); 1 EJ = 10^{18} J]. By comparison, the 2011 Tōhoku earthquake and tsunami in Japan had approximately 1.41 EJ of energy. The Eltanin impact equated to approximately 1.6 × 10^{10} kPa of pressure across an area of approximately 4000 km^3, or approximately 1,000 km away from the impact zone. This figure obviously dropped off at greater distances from the impact point, but basically it generated a massive pressure wave moving through the Pacific Ocean. Why is this important? Many people have heard of dynamite fishing, and it has been shown in the movies from time to time, such as in *Crocodile Dundee II* when Crocodile Dundee went dynamite fishing in Central Park, New York. When dynamite explodes, it generates a pressure wave that kills fish in the immediate area. Their swim bladders are ruptured, and they die and float to the surface. It is an unpleasant, illegal, but very effective fishing technique. For marine species without swim bladders (e.g., whales), death comes from ruptured ear drums, ruptured blood vessels, or brain damage. It is often a grim, lingering death over a few days as their entire navigation system shuts down.

The amount of pressure needed to this is approximately 57 kPa for a whale and 30 kPa for fish—much less pressure than the kind of force unleashed by Eltanin. The prognosis for anything in the water throughout the entire South Pacific Ocean, if not the entire Pacific Ocean, was not good. The story is grim: an initial wave 2.5 miles (4 km) high, a massive pressure wave moving through the Pacific Ocean, and a pulverized asteroid blasting material up into the atmosphere.

The good news, if there was any, is that the tsunami became relatively small quite quickly as it radiated away from the point of impact. Although it may have been as much as 2.5 miles (4 km) high at the point of impact, recent models show that it was probably "only" 65–100 ft (20–30 m) high by the time it hit Chile's coast some 9,000 miles (14,500 km) away.

Phew, "only" 65–100 ft high at the coast. To put this in context, the 2004 Indian Ocean tsunami that killed nearly a quarter of a million people was up to approximately 16 feet (5 m) high at the coast. Fortunately, if that is the right word, humans were not around the Pacific 2.5 million years ago, and so it was only everything else that suffered. Strangely, though, evidence for a massive tsunami inundating Pacific coasts around this time is at best rare and at worst nonexistent, depending on what one reads. As such, this chapter is a little disingenuous because while the tsunami may not have changed the world in any major way, it was by far the largest one we know of right now—at the point of impact anyway.

However, the effects of the other environmental disturbances associated with the tsunami were most probably far more devastating. Let us keep going and see what happened.

First, as a result of the pressure wave, the Pacific Ocean would have been littered with dead fish and marine mammals such as whales. What evidence is there for this? If we look at present-day situations in which marine mammals are sick or disorientated, we often see mass strandings and increasingly we see people desperately trying to return them to the water in a rush against time as the creatures overheat and die. Put this on a mega-scale, multiply up these mass standings to something unprecedented in history. Let them get washed up on the coast and lie there for more than 2 million years. As a result of all of this, one might expect to find thick layers of bones left behind at the coast. It is true that many fish and marine mammals would have sunk into deep water and been recycled by predators and scavengers, but not all. Is there any evidence for such a catastrophic die-off?

Bone Beds

Here, we must digress for a moment to take a quick journey through part of the geological timescale. As mentioned previously, the Quaternary started approximately 2.5 million years ago (2.58 actually). This was preceded by the Neogene period that ran from approximately 23.03 to 2.58 million years ago, part of which was made up of the Miocene epoch (from 23.03 to 5.333 million years ago). These get a brief mention here simply because several academic papers refer to them with regard to some unusual bone beds, although a recent paper written by one of this book's authors (JG) re-evaluated the ages and showed that they could just as easily be dated to 2.5/2.6 million years ago—the beginning of the Quaternary.

Mass strandings in recent times have generally been attributed to a variety of causes. Human activity in some way, shape, or form is viewed as a major cause, although this was not really important 2.58 million years ago. Other causes include herding behavior, large-scale oceanographic changes, and harmful algal blooms. However, these mass strandings are usually species specific. In other words, they will relate to a particular type of dolphin or whale or whatever.

If we assume that these unusual bone beds discussed in the academic literature do actually date to the correct time for this story, then there is an unusually dense accumulation of fossil marine vertebrates along the Atacama coast of northern Chile, where the bones of rorqual whales, sperm whales, seals, aquatic sloths, walrus-whales, and predatory bony fish are preserved.

This is not a normal mass stranding as seen today. They seem to be accumulated over a series of layers—a recurring process, in other words . . . storms driving the dead bodies in to shore? We do not know. What we do know is that this was not the only strange thing going on at that time. There was a marked change in seal populations. The Chilean coast went from being dominated by "true seals" everywhere to having only fur seals and sea lions in the north—why? We don't know. But, similar events also happened in South Africa, and possibly Australia and New Zealand as well. Did some survive better than others?

Interestingly, these bone beds also contain sharks, rays, seabirds, crocodiles, penguins, boney fish, gastropods, bivalves, and crustaceans. The bird bones seem to have been badly decomposed, and all seem to have traveled a long way. Maybe birds were killed closer to the asteroid impact zone? Needless to say, there would also have been a shockwave through the air as well as the ocean. Also, the asteroid would have been pulverized when it hit the ocean and in doing so would have thrown a lot of material up into the atmosphere. The effect on bird life is one possible indication of this. Let us now deal with the fallout . . . literally.

Unleashing Hell in the Atmosphere

Earth's atmosphere consists of a number of layers that differ in properties. The lowest layer is the troposphere, which extends from the surface to the bottom of the stratosphere. Three-fourths of the atmosphere's mass resides within the troposphere. This is the layer within which the Earth's terrestrial weather develops. The thickness of this layer varies between 10 miles (17 km) at the equator to 4 miles (7 km) at the poles. The stratosphere extends from the top of the troposphere to the bottom of the mesosphere. It contains the ozone layer. We all know that one.

If a 0.6-mile (1-km)-diameter asteroid (multiply these figures by 4 as a rough guide to how bad it could be if it is not bad enough already) strikes the ocean, there will be something in the region of 2.3×10^{10} kg of asteroid fragments, 4.2×10^{13} kg of water, 1.2×10^{13} kg of water vapor—all thrown up 9 miles (15 km) or higher into the atmosphere—that is a lot of stuff. The environmental outcomes of such disturbance could result in the annual mean temperature decreasing by 20 degrees Celsius in the lower stratosphere, and water (and sulfur) injected into the stratosphere would have long-term and global weather and climate consequences. Large masses of extra sulfur introduced into the stratosphere would be converted into sulfuric acid rain.

The effects of a moderately large deep-ocean impact event such as Eltanin may fall somewhere between that of a Mount Pinatubo eruption (the massive eruption of 1815 led to what is now known as "the year without a summer" or, more technically, a multiyear drop of global temperatures of at least 0.5°C) and the Chicxulub impact (this led to a 2°C drop in global temperatures for 3 years or longer—"a world without heat"). Furthermore, any excess water thrown up and not used to produce sulfuric acid rain may condense as water ice, increasing planetary albedo, causing the Sun's energy and heat to be reflected away from the Earth before it can reach the surface. This water ice would also interact with any veil of atmospheric dust generated by the event, thus making the loss of the Sun's energy and heat even worse.

However, all is not lost. The latest computer models suggest that if Eltanin fell where it did, then the effects would probably not last long, maybe a matter of a decade or so at worst, and certainly not enough to cause widespread glaciation and the start of the ice ages. Eltanin was not responsible for the massive global climate change that saw the onset of Northern Hemisphere glaciations; it fell too far south in the Pacific Ocean to have a global effect.

If this is a chapter about a mega-tsunami and all the associated environmental changes that fundamentally changed the world, then it is not delivering the final punch.

The Brotherly Wrinkle

There is a possible wrinkle in this "it wasn't that bad" scenario. As you may recall, Chicxulub probably had a smaller brother (Boltysh) that hit around the same time—65 or 66 million years ago. We know that large asteroids don't always strike the Earth by themselves. Large rocks hurtling toward the Earth have an annoying habit of breaking into bits. This has happened throughout geological history, but we always seem to get lost in Chicxulub and the demise of the dinosaurs. For example, there were probably three asteroids that hit around the Jurassic–Cretaceous boundary (~145 million years ago; they caused the extinction of ~20% of all genera) and also three at the end of the Devonian (365 million years ago or thereabouts; they caused the extinction of ~30% of all genera). There were others as well, all having similar catastrophic impacts on life on Earth.

Did Eltanin have a sibling? That is a difficult question to answer, but if we are suggesting that there is a link between Eltanin, the start of Northern

Hemisphere glaciations, and *H. erectus* moving out of Africa, then we need to at least have a look around because one or more siblings are needed to have struck the Earth in order to cause the climate change we see.

The clues are tantalizing, if a little vague.

There is a group of Chinese publications that wave the flag for a sibling asteroid. Like the Eltanin, some ship-based research produced a sediment core taken from the deep seafloor in the northern Pacific Ocean just to the northeast of the Marshall Islands [or ~2,175 miles (3,500 km) southwest of the Hawaiian Islands] that showed the traditional signals of an asteroid impact—geochemical markers (iridium) and solidified molten rocks (tektites). The same signals are replicated on land between several sites approximately 1,200 miles (2,000 km) apart in western, northern, and southern China, and then there are marked climatic changes noted at approximately the same time in both southern Japan and northern China.

If we add the possible northern Pacific sibling into the mix, then we can disturb the atmosphere in both the Northern Hemisphere and the Southern Hemisphere, with serious global climatic repercussions. Although the megatsunami generated by Eltanin (and its sibling? No one has looked) may have had little effect on humans, the related rapid climate change certainly caused a stir. Our ancestors eventually responded and moved out of Africa, and it all began from there.

Although we don't know enough yet, what we do know is worrying. And just to finish, scientists are pretty sure that all existing asteroids larger than 6 miles (10 km) in diameter have been discovered among the current Near Earth Objects population, and so the risk of a giant asteroid impact may be excluded for *at least the next century*. Phew.

However, there are probably well over 100 objects 0.6–1.2 miles (1–2 km) in diameter, and possibly thousands between approximately 0.4 miles (600 m) and 0.6 miles (1 km) in diameter, orbiting in the Earth's neighborhood that are as yet undiscovered. That gets us pretty much into the Eltanin range.

Afterword

As I (JG) was writing a later chapter for this book, I came across another possible link in the chain of post-impact events. A paper passed in front of me as I was researching Chapter 13 about a large submarine landslide off the coast of Norway. It is a bit of a spoiler, but most really large submarine landslides have some trigger event such as an earthquake or rapid sea level change, and the largest one that we know about was the Agulhas Slide off the coast of South

Africa that happened approximately 2.6 million years ago. The Eltanin impact undoubtedly caused an earthquake, but let's not forget the pressure wave as well. Hmmmm.

Fancy that.

And because we have managed to drift away from earthquakes for a while, let's keep going. After all, there are many ways to generate tsunamis, and we have been quite remiss in not really mentioning volcanoes yet, other than in passing comments. They have generated some of the more charismatic events in history.

11
Saved by the Baguette

> For they say that Poseidon inundated the greater part of the country because Inachus and his assessors decided that the land belonged to Hera and not to him. Now it was Hera who induced Poseidon to send the sea back, but the Argives made a sanctuary to Poseidon Prosclystius (flooder) at the spot where the tide ebbed.
>
> **—Section 2.22.4: writings of Paulinas, a Greek traveler of the 2nd century AD (cited in Jones, 1918)**

Who Cares?

It could be argued that because volcanoes represent only 4% of all historic tsunamis, there is a far greater need for researchers to focus their attention on those generated by undersea earthquakes. There are a number of ripostes that can be brought against this observation, not the least of which may be two names, Santorini and Krakatau. Both in their own way have left unique records of the destruction they wrought, and they continue to occupy the minds of scientists to this day.

Santorini (Thera)

Volcanoes are indeed an important cause of catastrophic tsunamis, none more so perhaps than Santorini (Thera), which erupted with devastating effect in approximately 1600 BC. This Late Bronze Age eruption generated a tsunami that rampaged throughout the Mediterranean, saw the beginning of the end for the Minoan civilization, and changed the entire political setup of the region at that time. Not bad for a volcano, a sort of volcanic coup.

Thera (well, we are calling it Thera—this is basically the island of Santorini) erupted and blew a massive hole in the island of Santorini in the eastern Mediterranean. This was a powerful eruption, but that does not tell us much. As a relative measure of its explosiveness, we use what is called the Volcanic

Explosivity Index (VEI). It had a VEI of approximately 7. This means that is was big. Although this scale of measure is actually open-ended, the largest volcanic eruptions that we know of have been given a VEI of 8, so Thera is right up there. Any volcanic eruption that has a VEI at the upper end of the scale between 6 and 8 is classified as "ultra-Plinian."

Plinian eruptions eject massive amounts of pumice, debris, and hot gases high into the stratosphere in a manner similar to that of Mount Vesuvius in 79 AD, which destroyed the ancient Roman cities of Herculaneum and Pompeii. Plinian eruptions are named after descriptions written by Pliny the Younger approximately 25 years after the event. Pliny's great attention to detail underpins our understanding of these Plinian events, and he even referred to the tsunami that was caused by the Vesuvius eruption: The "sea was being sucked backwards as if it were being pushed back by the shaking of the land." We all know about Vesuvius, and most people probably think that this was one of the biggest eruptions of all time. Well, no, it wasn't. It rates around about a VEI of 5, but that comes with a question mark after it because there is not really enough information to be more precise.

Ultra-Plinian eruptions are, not surprisingly, the big ones. They have ash plumes more than 16 miles (25 km) high, with volumes of erupted material between 2 miles^3 (10 km^3) and 200 miles^3 (1,000 km^3) in size. In being one of these, Thera is in exalted company with events such as the Lava Creek eruption of Yellowstone (circa 640,000 years ago; VEI 8), Lake Toba (circa 74,000 years ago; VEI 8), Tambora (1815; VEI 7), and Krakatau (1883; VEI 6). We discuss Krakatau (formerly known as Krakatoa until the spelling mistake was eventually corrected) later, but that is at the bottom of the ultra-Plinian scale, weighing in at a mere 2 miles^3 (10 km^3) of erupted material—Santorini may have been up to 10 times larger!

Given that Santorini was such a large event, it is somewhat of a minor distraction that geologists and archaeologists can't pin down the precise date on which it blew up. For this chapter, we have put it approximately 1600 BC, somewhere in the middle of the debate.

This catastrophic volcanic eruption and its ensuing tsunami may actually be behind the legend of the lost island civilization of Atlantis. The first written records that even mention Atlantis are in the *Timaeus* written by the Greek philosopher Plato in approximately 360 BC, more than 1,000 years after the actual event. To put that in context, it is a bit like us writing a detailed account of the English Battle of Hastings in 1066 AD (a mere 950 years or so ago) when the Normans invaded the country and having our work taken as being a true rendition of the events that unfolded. To be honest, even with the documentary evidence for 1066 AD and the famous Bayeux Tapestry that was created a few years later in the 1070s, we still don't even know the precise location of the battlefield.

In the *Timaeus*, Plato writes about a great island empire destroyed by earthquakes and floods, ultimately sinking into the sea. Although Plato is thought to have invented the details surrounding the story of Atlantis to suit his account of an ideal civilization, some scientists speculate that Atlantis was actually the island of Thera and the empire that of the Minoans based on nearby Crete a mere 75 miles (120 km) to the south, but as with most of these stories, things are a little more complicated.

Another early writer, Paulinus, commented on the effects of Santorini's eruption, specifically the tsunami. The quote at the beginning of this chapter is thought to be referring to the flooding of the plain of Argos [170 miles (270 km) northwest of Santorini] by the tsunami. The Argives in the quote are residents of Argos, and so this must have been a significant event. They placed the sanctuary at the maximum point of inundation; a similar practice has been reported for Japanese shrines and Māori Pā sites in New Zealand.

Returning to the actual event, it is thought that during the early stages of the eruption the volcano collapsed, allowing a huge amount of seawater to enter the magma chamber. When water comes into contact with hot magma, it is instantly turned into steam and produces a tremendous explosion—a phreatomagmatic eruption—which in this case also generated prodigious tsunami waves that could have been 200–300 ft (60–90 m) high. There is evidence that giant destructive waves inundated Crete, western Cyprus, and all along the eastern coast of the Mediterranean as far as Tel Aviv and Jaffa. There is also geologic evidence of a wave washing more than 200 miles up the Nile Valley at approximately this time. Many Mediterranean cultures, including those of the Egyptians, Babylonians, Greeks, and Hebrews, have flood myths that may have been prompted by the Santorini tsunami waves. There is even speculation that it was a tsunami which gave rise to the story of the parting of the Red Sea when Moses and the Israelites were fleeing the Egyptians. The withdrawal of water ahead of a giant tsunami wave would have exposed the seafloor and allowed the Israelites to cross, and then the following crest, coming in as a giant flood wave, would have crushed the pursuing Egyptian army. At any rate, the tsunami generated by the eruption of Santorini certainly ranks as one of the most important tsunamis to have occurred in the Mediterranean Sea.

It appears to have been responsible for the destruction of the Minoan civilization on Crete, and many papers have wisely stated this "fact." Unfortunately, these are murky waters with glorious flights of fancy, including Plato's undoubtedly mythological Atlantis catastrophe. Then there is the Greek myth of Deucalion's flood, which many Christian scholars continue to accept at face value, thus asserting that this occurred a few centuries later than the global one survived by Noah and company and variously dated to approximately 1530 BC. There are also the biblical seven plagues of Egypt and so on. What really happened at that time is best found in the physical evidence that is left behind.

We know from recent work that the tsunami generated by Santorini inundated the coast of Crete. The tsunami completely inundated the Minoan town of Palaikastro at the eastern end of Crete and effectively brought to an end the use of the palace at Malia on the north coast.

Palaikastro is one of Crete's many tourist beaches. (Figure 11.1-see color plate section) shows the small back beach cliff that is largely composed of the Santorini tsunami deposit. Bruins and co-authors wrote an excellent paper about this in 2008—well worth a read. Amusingly, many years ago, at a time before science came knocking and a fascination with all things tsunami-related began, I (JG) did exactly what these tourists are doing—sat on the beach sunbathing. I read the entire *Lord of the Rings* trilogy blissfully unaware that above my head was the most staggering example of a tsunami deposit. Unfortunately, I have not been back . . . yet.

The tsunami was a devastating blow to Minoan civilization, but it did not cause the end of it. Not surprisingly, there was a building boom in the aftermath of the tsunami. However, the major population centers of the Minoan seafaring power base that dominated the Mediterranean at this time were unwalled and most were located along the coastline. Although this seems entirely logical, this coastal focus would eventually lend itself to a slow lingering death for the Minoan civilization. The tsunami destroyed the mighty Minoan fleet and any coastal defenses that there might have been. Add to this the effects of the eruption that caused a global drop in temperatures leading to several years of cold, wet summers in the region that ruined harvests and you have the gradual decline of a once great people. Within approximately 50 years, it was gone, invaded and controlled by the military-minded authority of the Mycenaean Greeks, whose navy had been protected from the vagaries of Santorini's wrath by the harbors in the protected Gulf of Corinth.

The eruption of Santorini and its region-wide tsunami caused a sea change in the Mediterranean. Just like we see with contemporary society, most people expect tsunamis to be generated by earthquakes. If you are reasonably prepared, then you are aware that a tsunami might follow a large earthquake. However, when the unexpected happens and a tsunami comes out of the blue, it can have a devastating effect on a community's resilience, resolve, and social cohesion. Mediterranean societies were ill-prepared, but those that were fortuitously sheltered from the most damaging effects took full advantage of the power vacuum that followed.

Krakatau

Krakatoa, East of Java is a classic, 1969 US disaster film starring Maximilian Schell. The storyline is loosely—very loosely—based on the 1883 eruption of Krakatau (for a long time, "Krakatoa" was the accepted English spelling, probably due to a typo in the reporting of the 1883 event). Clearly, geography meant nothing to Hollywood. Although Krakatau is west of Java, the producers believed that "east" conjured up more atmospheric images of the Far East. And so for years it provided teachers with an excellent way of helping their acolytes come to grips with the geography of Indonesia.

The Krakatau eruption is commonly called the largest volcanic tsunami in historic times and also the most famous historically documented one at that. It is probably the latter that leads us to say the former because we tend to forget the Mount Tambora eruption of 1815. This had a VEI of 7, as opposed to a VEI of 6 for Krakatau. It was well over 10 times larger, and during the eruption Tambora's peak shrank from 14,100 ft (4,300 m) to just 9,350 ft (2,850 m). The Tambora eruption led to such extreme climatic disruptions that the next year, 1816, became known as the "year without a summer." The Northern Hemisphere faced famine, crop failures, and widespread death of livestock. However, it only generated a small, local tsunami, up to a maximum of 13 ft (4 m) high. This was largely because the eruption caused little disturbance of the sea.

And so to Krakatau. Krakatau is a different beast. Like Tambora, it is in the Indonesian archipelago, one of the most geologically active zones on Earth. Major earthquakes occur along the Java and Sunda Trenches that form the southern and western margins of the archipelago, respectively, and many of the Indonesian islands "behind" these trenches were built by explosive volcanoes.

The most well-known recent earthquake on the southern margin was the 2006 Java Trench event offshore from Pangandaran, southwest Java. This generated a 6-ft (2-m) tsunami that inundated Shark Bay, northwest Australia, some 1,000 miles (1,600 km) away. As for the most well-known earthquake to the west, that is undoubtedly the 2004 Indian Ocean earthquake and tsunami, which are discussed in Chapter 15.

Not surprisingly, Indonesia has many stories of volcanoes, earthquakes, and tsunamis. For example, the Pararaton, or Javanese Book of Kings, completed sometime around the 15th or 16th century, notes a massive eruption in the year 338 Saka (416):

> A great glowing fire, which reached the sky . . . the noise was fearful, at last the mountain Kapi [thought to be Krakatau] with a tremendous roar burst into pieces and sank into the deepest of the earth. The water of the sea rose and inundated the land, the country . . . was inundated by the sea; the inhabitants . . . were drowned and swept away with all property. . . . The water subsided but the land on which Kapi stood became sea, and Java and Sumatra were divided into two parts.

Geologically, the closest eruption of Krakatau that we know of around this date was in 535. As can often occur with what was undoubtedly an oral tradition later transcribed in writing, the message is the important point, not the details of time or often even place. However, we do not know what that message was or how it changed or was augmented through time. Thus, although we should not necessarily accept the story as being historically accurate, we should recognize that it is true to the message it is telling.

Whatever the case, this sounds like a tsunami generated by an ancient eruption of Krakatau. The reason for thinking this apart from anything else is that we know Krakatau generates tsunamis. It is the combination of massive eruptions coupled with the fact that the volcano sits in the Sunda Strait between Sumatra to the northwest and Java to the southeast. The volcano rises up directly from the seafloor, and unlike Tambora, which has some ground around it, any large eruption here will seriously upset the sea. Sunda Strait also has a configuration that helps magnify the effects of any waves. A narrow 15-mile (24-km) bottleneck at its northeastern end keeps the majority of any tsunami energy inside the strait in that direction. Furthermore, it is deep to the west and shallow to the east, allowing waves to vary greatly in height as they bounce around inside this confined space. In the 1600s, Sunda Strait was an important shipping route and gateway for early European colonizers of the region through to the incredibly important and jealously guarded Spice Islands of Indonesia sandwiched between the Molucca and Arafura Seas. However, because of its eccentric geography, Sunda Strait was also notoriously difficult to navigate. Perhaps this should have been a warning to keep away and try another route because Krakatau, peaceful and serene at the northern end of the strait, was about to wake up.

After sleeping for nearly 200 years, Perbuatan, one of the three volcanoes making up the island of Krakatau, suddenly awoke on May 20, 1883, with a series of explosions. Eruptions continued throughout the summer, creating vast fields of floating pumice, which reportedly spread far out across the Indian

Ocean. Then on August 26 and 27, the eruption came to a spectacular climax as Krakatau blew itself to pieces (Figure 11.2). The finale was so violent that it has been difficult for geologists to be sure exactly what happened, although a basic chronology has been put together.

The first major tsunamis were generated on the afternoon of August 26. At approximately 5:30 p.m., waves 3–6 ft (1–2 m) high struck the town of Anjer, 30 miles (50 km) away on the eastern side of the strait, breaking vessels loose from their moorings and damaging the town's drawbridge. Then at 7:30 p.m., a tsunami with waves 5 ft high surged into the town of Merak 12 miles (20 km)

Figure 11.2 An 1888 lithograph of the 1883 eruption of Krakatau.
Source: Report of the Krakatoa Committee of the Royal Society (1888).

farther north, smashing boats and washing away a Chinese camp. Throughout the night, small explosions from Krakatau continued to be heard and small tsunami waves periodically rolled ashore along the coasts of the Sunda Strait.

The next morning, the eruption became even more severe. At approximately 6:30 a.m., a tsunami wave estimated at 33 ft (10 m) high crashed ashore at Anjer. The town and harbor were demolished, and almost the entire population was killed. By approximately 7:30 a.m., the tsunami had surged into Merak, and as the Dutch engineer, Abell, fled, he looked back to see a "colossal wave" crash ashore. In the harbor at Telok Betong 50 miles (80 km) to the north in Sumatra, the Dutch warship, *Berouw*, was torn from her anchorage and cast onto the beach.

The eruption intensified, and by mid-morning the falling ash was so thick that the entire region was left in nearly total darkness. At 9:58 a.m., a cataclysmic blast occurred, throwing a column of ash up to a height of 82,000 ft (2.5 km). This explosion was heard as far north as Manila 3,700 miles (6,000 km) away, as far south as central Australia 2,500 miles (4,000 km) away, and as far as Sri Lanka and Rodrigues Island more than 3,000 miles (4,800 km) to the west. More than two-thirds of Krakatau disappeared in the explosion, including the entire volcanoes of Perbuatan and Danan and half of Rakata. Most of the material generated during the entire 3-month eruption was produced on this single morning. Huge pyroclastic flows poured down into the sea, some creating their own explosions. Thick deposits covered the sea floor extending for approximately 10 miles around Krakatau.

It was at this time that the largest tsunami was generated. But there were no witnesses to the event. The entire population of the island closest to Krakatau, Sebesi, a mere 8 miles (13 km) away, was annihilated. Even farther away, there were few who witnessed the largest tsunami come ashore—they had either been killed by earlier tsunami waves or fled into the hills. At 10:18 a.m., the giant tsunami struck Telok Betong, where it re-floated the beached vessel, *Berouw*, and carried her up a narrow valley, leaving the ship some 30 ft (9 m) above sea level, nearly 2 miles (3 km) inland. All 28 of her crew were killed. At 10:32 a.m., the giant wave swept over the ruins of Anjer and Merak, killing more than 10,000 people. Not a single building, not even the lighthouse, was left standing in Anjer. Merak, situated at the head of a funnel-shaped bay, may have experienced the greatest wave anywhere on the coast of Java. Here, the tsunami reached an incredible 135 ft (41 m) above sea level and tossed coral blocks weighing 100 tons up on the shore. At Princes Island off Java and at the lighthouse at Vlakke Hock, 50 miles (80 km) southwest on the eastern tip of Sumatra, the water was said to have reached a height of 50 ft (15 m). And at

Ketimbang, Sumatra, some 25 miles (40 km) northwest across the sea from Krakatau, the largest wave was 80 ft (24 m) high.

All in all, the tsunami waves had swept over nearly 300 coastal towns and villages. The official death toll listed 36,417 killed, but several thousand bodies were swept out to sea and never found. The true death toll will never be known, but it probably exceeded 40,000, almost all casualties of the tsunami waves as opposed to the eruption.

The tsunami traveled out from Krakatau across the Pacific and Indian Oceans. At approximately 9:00 p.m. that evening, the sea withdrew at Bombay, India. Overjoyed onlookers rushed out onto the exposed tidal flats to pick up stranded fish, narrowly escaping when the sea surged back in. On the northeast coast of Sri Lanka, four adults and three children were washed from a sandbar by a large wave. They were soon rescued by fishermen in boats, but one of the adults, a women, died 2 days later of her injuries. She was probably the most distant fatality produced by Krakatau; her death well illustrates how tsunamis have the ability to reach out over vast distances to inflict their destruction.

The tsunami was even recorded in the Atlantic Ocean. It was detected on tide gauges in the Bay of Biscay, and at 9:35 p.m. on the evening of August 28, a ½-inch tsunami wave registered on the tide gauge in the port at Le Havre, France, more than 10,000 miles (16,000 km) away.

The eruption was literally felt around the world as airborne shock waves repeatedly circled the globe. It became the "catastrophe of the century." A month after the eruption, corpses still littered the beaches of Java and the Sunda Straits. Pumice from Krakatau was reported at sea by ships for 2 years after the eruption; some of it reached as far as Natal in South Africa, more than 5,000 miles (8,000 km) away. More than 1 year later, 5-ft (1.5-m)-diameter trees, their roots jammed with pumice, washed ashore on Kosrae in Micronesia, nearly 4,000 miles (6,500 km) to the northeast. And almost 4,000 miles to the west, human skulls and bones were discovered 1 year later, washed up on a beach in Zanzibar. As an intriguing aside, these floating islands are often considered as one of the ways that flora and fauna get distributed around the vast expanses of ocean basins. This could explain how some species end up in the most remote island locations or how apparently human-introduced species such as rats arrived well before any humans did.

Just how were these devastating tsunamis generated? Geologists have proposed four processes that could account for the tsunamis, all four of which took place to one degree or another during the eruption: (1) the collapse of the huge mass of Krakatau Island into the sea; (2) submarine faulting around the margin of the caldera; (3) enormous quantities of volcanic material falling into the sea as pyroclastic flows; and (4) after submergence of the vents, explosions bursting upward to the ocean surface.

Some geologists believe that the explosion theory best fits the evidence. Early speculation that the eruption deposits are laid out in a concentric arrangement was confirmed by a recent underwater survey. This nondirectional distribution has been interpreted to mean that the deposits were laid down by a giant explosion, which could have produced the giant tsunami. Furthermore, at 15 different locations around Krakatau, the first sign of the giant tsunami was a rise in sea level, not a withdrawal. This indicates that the water was displaced upward at its point of origin by an explosion.

Other scientists are not quite so sure and point to an initial drop in sea level observed at some locations on the mainland coast. They note that seawater rushing in to fill the newly formed submarine cavity created by the subsidence of parts of Krakatau would account for this initial retreat of the sea, hence favoring collapse of the volcano as the mechanism generating the largest tsunami.

Yet another group of scientists favor "several cubic kms of pyroclastic flow material entering into the sea immediately after each of the large explosions" (Self and Rampino, 1981, p. 703) as the mechanism generating the tsunami waves. These flows may have had internal temperature of up to 1,000°F and could have traveled at a speed of more than 300 miles (480 km) per hour. It has even been suggested that as the flows spread away from Krakatau, the heavier parts of the flow would have sunk under water and continued along the bottom as a submarine flow, while the lighter parts—those less dense than water—would have roared across the surface of the water "cushioned like a hovercraft on their own escaping gases" to quote Krakatau expert, geologist Steve Self. And a pyroclastic flow can cause a tsunami even if it does not sink. Flows that crossed the Sunda Strait on the sea surface may well have caused a wave merely by pushing water out of their way. And there is evidence that pyroclastic flows from Krakatau actually did flow across the surface of the water, producing an incident known as the "Burning Ashes of Ketimbang."

Ketimbang was a village of approximately 3,000 inhabitants. During the eruption, the Dutch Controller and his family, the Beyerinks, had taken refuge in their hillside cabin, when the village was suddenly engulfed by darkness and then fire. Having shut all the doors and windows to the cabin, they saw burning ash begin fountaining up through the cracks in the floorboards. The family was badly burned, and one of the children died along with approximately 1,000 villagers incinerated by the incandescent ash. A pyroclastic flow had crossed the 25 miles of ocean from Krakatau.

Not all of the theories outlined previously are mutually exclusive. Many geologists seem to agree that the most likely explanation for the tsunamis prior to the 10:00 a.m. giant wave is that they were generated by pyroclastic flows. Many of the small tsunamis during the night were highly localized and could have been formed as material ejected from the volcano fell into the sea at various sites around the island. And the giant tsunami, generated at approximately the time of the culminating blast, could well have been created by the coincidence of rapid pyroclastic flows entering the sea, the sudden slumping of half of the Rakata cone into the actively forming caldera, and the gigantic explosion as cold seawater met molten magma.

In more recent times, the terror of the 1883 eruption has been revisited upon Sunda Strait and the inhabitants who live along its coastline. This all stemmed from the child of Krakatau. On December 29, 1927, Anak Krakatau, the "Child of Krakatoa," emerged from the collapsed hole or caldera that was formed after the full force of the 1883 eruption had destroyed its mother. The child grew rapidly, as much as 22 ft (6.8 m) per year, and in 2012 volcanologists warned that a tsunami caused by flank collapse of Anak Krakatau could occur. Then in 2018 the volcano started a new eruptive phase, including a strong Strombolian eruption in October. This is moderately mild in terms of volcanic eruptions, having a VEI of approximately 3, and it ejected lava bombs into the sea around it.

All of this activity culminated in the southwest sector of the volcano, including the summit, collapsing in an eruption on December 22, 2018. The volcano lost more than two-thirds of its volume, collapsing from 1,109 ft (338 m) to 360 ft (110 m) in height. This caused a tsunami up to 16 ft (5 m) high that, like the 1883 event, affected the coastlines of both Sumatra and Java. Unlike the 1883 event, however, this was recorded on video and reported throughout the world: 437 died and 14,059 were injured. Although it may have paled in significance in size to its mother's efforts, this is so far the deadliest volcanic eruption of the 21st century.

To demonstrate how much more we know about such events, as much as 6 years prior the event, scientists had considered the possibility of its occurrence, even correctly identifying the likely point of collapse. However, the problem we still face is that although we can definitely say that an event such as this will in all likelihood happen, we cannot predict in any useful time frame exactly when it will happen. Do you evacuate everybody? Do people take the information provided and self-evacuate? Do people give up their livelihoods and homes knowing that the volcano will undoubtedly erupt at some point and this could lead to a tsunami? Or do you just shrug and say "What will be, will be" and carry on enjoying the beauty of the environment, the security of your job, the fertility of the soil, and all of the benefits of living around these ticking time bombs?

And Now to the Baguette

We have wandered around the world in our look at volcanoes but have so far avoided the baguette. So let us now briefly visit Martinique. This island, along with Guadeloupe to the north, forms one of the five overseas departments and regions of France. It is located in the Lesser Antilles in the eastern Caribbean Sea between Dominica and St. Lucia. As islands go, it is reasonably large, with a land area of 436 square miles (1,128 km^2). In one of the world's many political quirks, as a region of France, it is part of the European Union.

In the northwest of the island sits the lovely Caribbean town of St. Pierre, founded in 1635. It was an important cultural and economic center known as the Paris of the Caribbean. Today, it is a popular tourist destination, visited by numerous cruise ships, but it would have been a very dangerous place to visit on May 8, 1902.

St. Pierre sits at the base of Mount Pelée, French for "bald" mountain, a name that sounds surprisingly similar to Madame Pele, the Hawaiian Goddess of Volcanoes. But unlike Madame Pele, who calmly pours out liquid magma, building islands of solid basalt rock, Mount Pelée has a fiery temper, exploding with rage. Explosive volcanoes, such as Mount Pelée and other Caribbean volcanoes, rapidly erupt, expelling burning hot gas and fiery ash—more like "fire and brimstone" than flowing rivers of molten lava. Explosive volcanoes can create pyroclastic flows, described by volcanologists as "incandescent, ground-hugging clouds, driven by gravity and fluidized by hot gases" (Francis and Self, 1983, p. 174). The May 8 eruption of Mount Pelée was extreme even for these violent events, and the phenomenon now bears the French name *nuées ardente* or "burning cloud"—a veritable volcanic avalanche of gas, dust, ash, incandescent solid particles, and fragments of lava.

Mount Pelée had already produced an explosive eruption back in 1792, but of course by 1902 there was no one around who remembered that event. But there were warning signs that began in early 1902 as a volcanic dome began growing near the summit of the volcano.

Then on May 5, a violent debris flow made up of pyroclastics, rocky debris, and water, called a lahar, slid down the side of the volcano into the ocean, creating a 13- to 16-ft (4- to 5-m) local tsunami, which killed approximately 100 people in St. Pierre. This was the main tsunami, but just like the asteroid discussed in Chapter 10, it was the associated hell that makes this event so bad. It was only 3 days later, just before 8:00 am on May 8, that the dome on Mount Pelée suddenly collapsed, creating a monstrous nuée ardente that roared into St. Pierre. It has been estimated that the nuée ardente traveled up to 85 miles per hour (140 kilometers per hour).

It completely destroyed St. Pierre.

The city was literally flattened. Nearly the entire population of some 29,000 people was almost instantly killed from suffocation and burns that scorched their skin and lungs, with temperatures estimated to be as high as 1,800°F (1,000°C). The nuée ardente continued on its path of destruction, flowing into the harbor, where it destroyed some 20 ships either by burning or by the tsunami waves it generated—not as big as the previous one but still devastating. There were no survivors from any of the ships. Indeed, there were only three survivors in the town. One was a shoemaker who lived on the outskirts of town, hence at the edge of the flow, and another was a young girl who had run, jumped into a boat, and fortunately been washed out to sea alive by the pressure wave in front of the nuée ardente as opposed to being burnt or drowned.

Perhaps the most amazing survival story, however, is that of Ludger Sylbaris, an Afro-Caribbean man who had been locked up in jail at the time of the eruption. There are different accounts of why he was in a dungeon-like jail cell. Some stories say he was involved in a bar fight, other accounts tell of him being drunk and causing a riot, but the story I (WD) was told when visiting the jail site in St. Pierre was that he had been thrown in jail for stealing a baguette from a local bakery.

In any case, it was the jail cell, partially cut into the mountainside and with its door facing away from the volcano, that saved his life (Figure 11.3). Although badly burned on his back, arms, and legs, Ludger survived and went on to become a celebrity, ultimately traveling with the Barnum & Bailey Circus.

Unfortunately, because geologists had no real grasp of volcanology in the early 20th century, no one thought to evacuate St. Pierre in the days before the eruption that caused the nuée ardente simply because such an event was beyond their comprehension. We now know so much more about these types of event and would term this "burning cloud" as a type of pyroclastic flow such as the one that occurred in 1980 when Mount St. Helens erupted. One question that remains to be answered, however, is what actually caused this to happen at Mount Pelée. There is still a good bit of scientific debate, but at least scientists now have a far better understanding of when these events can happen. We may not be able to say exactly when they will happen, but armed with scientific advice of an imminent threat, people are at least able to make informed decisions and, it is hoped, be safe.

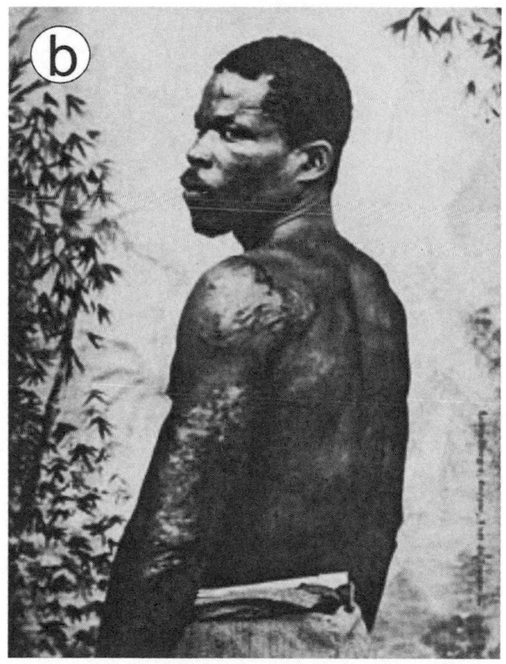

Figure 11.3 (a) The jail cell where Ludger Sylbaris survived the eruption. (b) A postcard showing Ludger Sylbaris, survivor of the huge Mt. Pelée eruption, Martinique, May 8, 1902, showing burns on his arm and back.
Sources: a, W. Dudley; b, unknown, public domain.

For Martinique, it is probably not surprising that historical records show that the island is mostly subjected to tsunamis originating from either itself or some of its near neighbors, such as St. Vincent or Guadeloupe. The islands all have similar origins and similar ways of generating tsunamis. But every now and again, they get one that comes out of left field, or perhaps in this case we should say right field because the only tsunami to affect the island historically from the east is the Lisbon tsunami of 1755. It caught not only the islanders by surprise but also those much closer to the source.

12

1755, Lisbon

The Benefit of Brothels

> The extraordinary commotions of the sea . . . having happened within a few hours of the great earthquakes; one which shook Spain and Portugal . . . will naturally be ascribed by everybody to those earthquakes. . . . Now for my part I can hardly persuade myself, that the bare agitations of the earth, at those times, could be great enough, to put the sea into such vehement commotions, as it appears to have been in. . . . To account for these things satisfactorily, it seems to me that we must have recourse to such an eruption of the vapors which caused those earthquakes . . . these furious vapors impatient of restraint, must have continued to drive along through their subterraneous passage, till they found some place, where the top of the caverns, which contained them, was not of sufficient strength to confine them; and there they would burst out of their dungeons, and spring up into day. The eruptions which caused those uncommon motions of the sea that surprised the inhabitants of . . . St. Martin's, were very probably made in the Atlantic ocean, to the eastward of the West-India islands, and near the same latitudes.
>
> **—Winthrop (1755)**

A Case of the Vapors

As scientists, it is very easy for us to sit in the existing bubble of our knowledge and look smugly back at the fanciful explanations given over the past centuries for natural events that we understand far better today—and in 250 years' time there will doubtless be smug scientists looking back at our somewhat rudimentary understanding of things too.

In this modern world, it is difficult for us to realize that what we know as earthquakes and tsunamis were still part of the giant scientific melting pot of the 17th and 18th centuries, a time when geology was not yet a clearly defined discipline and explanations for some of the Earth's features were just starting to be expressed as occurring over geological and not biblical timescales. Throw into this melting pot the instantaneous and often catastrophic events that were happening at the time and there were, not surprisingly, numerous theories up for grabs. One of the most popular theories at the time, as expressed by Professor Winthrop in the opening quotation, was that vapors trapped underground in vast caverns would on occasion burst forth from their captivity and in doing so cause earthquakes and, particularly if they occurred under the sea, generate large waves that would devastate coastal communities. In trying to come to grips with tsunamis that affected both the Caribbean and western Europe, Winthrop essentially conjures up two sets of vapors bursting out from their subterranean dungeons—one lot close to Portugal and Spain and the other somewhere just to the east of the Lesser Antilles.

So what did happen, and how much wiser are we today?

All Saints' Day

November 1, 1755, was not a good day to be in Lisbon. It was not a good day to be anywhere around the coastlines of Portugal, Spain, Morocco, and all of the Atlantic coast on either side of the Strait of Gibraltar and much, much farther afield as well.

On the morning of All Saints' Day, many Lisboetas—people of Lisbon—were packed into the churches of the city attending religious celebrations when a powerful earthquake with a magnitude recently estimated at approximately 8.5 ± 0.3 violently shook the western and southern areas of the country at around 10.00 a.m. It is still one of the most violent and longest seismic events on record, lasting something in the region of 9 minutes. The earthquake epicenter was located approximately 125 miles (200 km) southwest of the southwestern tip of Portugal along the Azores–Gibraltar fault zone, which is the westernmost continuation of the boundary between the Africa and Eurasia plates. Although this sounds wonderfully precise, neither the location nor the character of this plate boundary are particularly well understood, and in reality the precise source of the 1755 earthquake remains unclear.

The quake was so strong that its effects were observed as far away as England, where Viscount Parker, F.R.S., noticed a disturbance of the water in the southwest corner of his moat around Shirburn Castle in Oxfordshire. The Viscount's observations, along with many others, were discussed over the ensuing months at meetings of the Royal Society in London, the oldest scientific society in the world. The president at the time, George Parker, 2nd Earl of Macclesfield, was Viscount Parker's father. At the meeting on Thursday, February 26, 1756, well after his son's observations had been discussed, they read out a letter from Father Joseph Stepling of Prague (in the Czech Republic) to James Short, a Fellow of the Royal Society. In the letter, Stepling recounted the effects of the earthquake on the hot spring at Toplitz (a beautiful spa town even today) a few hundred miles farther away from Lisbon than Oxfordshire. Here, the natural hot spring water supply started to behave erratically, but after a while it settling down again—the quality of the spring apparently improved, as did the flow, with the baths being supplied with more hot water than usual and with a higher mineral content. There were strange things afoot in the bowels of the Earth, but things were far worse in Portugal.

In Lisbon, church roofs collapsed, and those not crushed by the debris ran out into the streets. Fires started as votive candles and lamps fell over during the shaking, and people sought refuge along the banks of the Tagus River and around the city's harbor front. All of this took some time, and so just as the panicked masses started to feel marginally safer nearer the water and were watching the fiery disaster unfold, they were struck by a tsunami rushing in from the Atlantic Ocean.

Actually, in Lisbon the sea withdrew first as the negative wave arrived (just like any wave, a tsunami has a peak and a trough—when the trough of a tsunami arrives at the coast first as it often does, then the water withdraws, often well below any level that people have seen before; this is a valuable, and possibly the last, useful warning sign for people to get inland and high up as soon as possible). Then a great wave came roaring in, penetrating more than half a mile into the city, rushing up streets and inundating houses. Bridges were broken, walls overturned, and great piles of debris swept away and carried off shore. At Belem Castle, the wave was estimated to have been as high as 50 ft (15 m). One ship's captain said that a wave at least "20 feet [6 m] high rushed over the quay immediately after the shock." The ships in the Tagus River were said to have been "tossed about as in a violent storm," leaving the harbor a "forest of entangled masts." Many thousands of those people who had survived the earthquake were swept away and drowned by the tsunami (Figure 12.1).

Lisbon may have suffered the full impact of the tsunami, but it was not the only place to suffer. The first wave arrived at the city of Cadiz, some 200 miles (350 km) southwest of Lisbon and close to the Strait of Gibraltar, approximately 1 hour after the shock and was said to have been 60 ft (18 m) high. It tore away portions of the town wall, which weighed 8–10 tons. It washed over the Valdelagrana Spit in the Bay of Cadiz, an area now described as a "wide, lengthy beach with calm surf, water activities and a promenade with restaurants, bars and shops." It was not quite so calm in 1755. Furthermore,

Figure 12.1 Lisbon, Portugal, during the great earthquake of November 1, 1755 (copper engraving from 1755 showing the city in ruins and in flames).
Source: The Earthquake Engineering Online Archive—Jan Kozak Collection: KZ128 (http://nisee.berkeley.edu/elibrary/Image/KZ128).

in Roman times in approximately 218 BC, it was also overtopped by the predecessor to the 1755 tsunami. Studies offshore have shown that these major events seem to occur approximately once every 1,500–2,000 years, but there is no information on smaller ones or those that originate from other sources; these may not be as deadly, but they can kill just as well. Unfortunately, these small ones tend to sneak under the geological radar largely because every time an event such as the 1755 tsunami occurs, it often wipes the geological slate clean of any evidence related to their smaller cousins. The message here is not to be complacent—enjoy the sand, the bars, and the shops, but be aware, not scared, that anything can happen at the coast.

Back to 1755. At Tangier, Morocco, on the other side of the Strait of Gibraltar, the first wave is said to have been 50 ft (15 m) high and to have flowed 1½ miles (2½ km) inland. But the tsunami also propagated across the Atlantic Ocean. Two hours after the quake, a wave estimated at more than 18 ft (5 m) in height struck Madeira in the Azores. Similar waves were reported from both the Canary Islands and the other islands of the Azores. Tsunami waves traveled north to the British Isles, where in Wales a large "head of water" rushed up the river at Swansea, breaking the mooring lines of vessels, casting them adrift. Between 2:00 and 3:00 p.m., a large mass of water poured into the harbor near Cork in Ireland, rising more than 5 ft (1.5 m) over the quay.

Galway, on the west coast of Ireland, was also hit, resulting in partial destruction of the "Spanish Arch" section of the city wall.

William Borlase, a vicar in Cornwall, a polymath, gentleman scientist, Cornish antiquary, geologist, and naturalist, writing in 1758 in his *The Natural History of Cornwall* recorded the effects of the tsunami in the region. At approximately 2 p.m. that day, some 4 hours after the earthquake things started to go awry. In harking back to the words of Professor Winthrop, Borlase stated,

> What connection with or relation to these violent convulsions on the Continent our little, and (thanks to Providence) momentary agitations of the Sea on the coasts of Britain had, 'tis difficult to say; but their happening both on the same day, and within a few hours of one another, the many repeated fluctuations in the river Tagus as here in Cornwall, by alternate swells and sinkings, the shocks felt on the same day far to the West by several ships; all these circumstances seem to declare very consistently, that what we felt was either the fainter parts of that deplorable shock at Lisbon, or the last expiring efforts of some subterranean struggles further to the West and South-West under the Atlantic Ocean. . . .Indeed, it can scarce be imagined, that a shock, so far off the coast of Spain, could be so immense as to propagate so violent a motion of the water quite home to the shores of Britain in less than five hours; I should rather think that the same cause, diffused in different portions through the intestines of the earth, produced several subsequent rarefications of the imprisoned vapors; that these rarefied tumid vapors affected the Seas and land above them in proportion to their own power, the dimensions of the caverns they had to extend themselves in, and the superior or fainter resistance of the incumbent pressure.

The tsunami also crossed the Atlantic Ocean to the west and struck the islands of the Caribbean. In a veritable who's who of Caribbean islands, it reached estimated heights of 6–20 ft (2–6 m) on Anegada and 12 ft (3.5 m) on Antigua; it was reported in Hispaniola, Cuba, Martinique, and St. Martin; and it flooded streets and wharves in Barbados [~3,600 miles (5,800 km) from Lisbon]. It was not noticed along the US Atlantic seaboard, which is strange because recent modeling of the event shows the coastline being affected. We will probably have to wait until some early documents are unearthed, but this does tell us something useful. We should never believe that we know everything about these old events simply because they were recorded in writing at the time. It may have all been written down at the time, but our knowledge is only as good as our ability to find those writings. Did the tsunami inundate the US eastern seaboard? It probably did, but we cannot be sure.

As a case in point, a computer model produced in the early 2000s of the tsunami waves as they spread out across the Atlantic Ocean from where we think the earthquake happened shows that the energy of these waves was focused in three main directions: (1) the northeastern areas of North America—Newfoundland, Nova Scotia, and probably down the coast to Boston; (2) the Caribbean (we know about that one); and (3) Brazil. Well, guess what? As recently as 2015, it was discovered that the tsunami waves may have reached the coast of Brazil, then a colony of Portugal—the computer model showed it before we even found the evidence for it. Letters sent by Brazilian authorities at the time of the earthquake describe damage and destruction caused by gigantic waves (one wonders what we might find for North America).

This impact on Brazil is very important because Lisbon was the main recipient of the immense wealth in gold and diamonds that was being produced by Brazil. The widespread destruction that happened in Lisbon in 1755 took place in one of the richest and most opulent cities of 18th-century Europe. The losses were catastrophic, not just to people and property but also economically. It caused a loss equivalent to 8–12% of the Portuguese capital stock (this is about twice as much economic damage as was caused by the 2011 tsunami in Japan). To put this in an 18th-/19th-century context, when Portugal finally lost Brazil a few decades later in the 19th century, the economic loss to Portugal was of a similar magnitude, but that came about after a long drawn-out series of events caused by war and politics—the end was inevitable. In 1755, this was instantaneous and devastating.

Destruction and Rebirth

In Lisbon, the official death toll is usually placed between 10,000 and 100,000 people. Outside Lisbon, approximately 2,500–5,000 people died, including all of those from the earthquake, fires, and tsunami in Portugal, Spain, and Morocco. Virtually all of the large buildings in Lisbon were destroyed. In a staunchly Catholic country, it was devastating: 32 churches, 31 monasteries, 75 nunneries, and 60 small chapels.

Although 33 palaces were also destroyed, amazingly the royal family escaped unharmed from the catastrophe. Fortunately for them, King Joseph I of Portugal and the court had left the city after attending mass at sunrise and fulfilling the wish of one of the king's daughters to spend the holiday (All Saint's Day) away from Lisbon.

Some 250 aftershocks occurred in the 6 months after the main earthquake, which led to the collapse of many unstable buildings. Joseph I developed a fear of living within walls. As a result, the King's court was accommodated in a huge complex of tents and pavilions in the hills of Ajuda, then on the outskirts of Lisbon. Like the King, the Prime Minister, Sebastião de Melo (1st Marquis of Pombal), also survived the earthquake.

After the earthquake, the King turned increasingly to religious activities, losing interest in the day-to-day affairs of state. Pombal stepped into this power vacuum with great skill and effectively took control of all governmental operations. The nobles and the Jesuits who perceived him as a threat to their positions of influence failed to unseat him, thus effectively giving him absolute control over the Portuguese nation. In a sense, this is just what was needed—his vision for Lisbon resulted in the beautiful city we see today, but it required the sometimes brutal hand of a near-dictator to achieve.

When he was asked what was to be done, Pombal reportedly replied "Bury the dead and heal the living," and he set about organizing relief and rehabilitation efforts. Firefighters were sent to extinguish the raging flames, and teams of workers and citizens were ordered to remove the thousands of corpses before disease could spread. The figures for the number of dead are not very accurate, mainly because there was no reliable estimate of the Portuguese population before 1755. But also, contrary to custom and against the wishes of the Church, Pombal made sure that corpses were disposed of as quickly as possible. Due to fears of plague and disease, bodies were either placed in mass graves or loaded onto barges and buried at sea beyond the mouth of the Tagus. To prevent disorder in the ruined city, the Portuguese Army was deployed, and gallows were constructed at high points around the city to deter looters; more than 30 people were publicly executed. The army also prevented many able-bodied citizens from fleeing, pressing them into relief and reconstruction work.

Things moved fast.

Pombal assembled a team of designers to guide the rebuilding of the city with the aged 83-year-old engineer-in-chief, General Manuel da Maia, in charge. He was well-grounded in the problems that were facing the city but was also respectful of traditional building styles. On December 4, 1755, little more than 1 month after the earthquake, he presented his plans for the rebuilding of Lisbon. Maia presented options ranging from abandoning Lisbon to building a completely new city. The "clean slate" option was to create a new city. However, this option had several negative features, including the demolition of existing buildings (among these were, ironically, Pombal's home, the one Protestant Church still remaining, and the brothels—we return to these later). Complete destruction of the surviving buildings would have occurred at a time when housing was critical and conditions would have become worse before they improved. It would have created a legal nightmare with regard to property rights and would have sent a devastating psychological message to the people still living in the city that they were being abandoned.

Pombal decided not to go with this option—whether out of altruism or self-interest is uncertain.

The second and third plans proposed widening certain streets. The fourth option boldly proposed razing the entire Baixa quarter—the center of Lisbon—and "laying out new streets without restraint." This last option was a happy medium and was, not surprisingly, chosen by the King and his minister. In less than 1 year, the city was cleared of debris. Keen to have a new and perfectly ordered city, the King (Pombal really) commissioned the construction of large squares, rectilinear, large avenues, and widened streets—the new *mottos* of Lisbon. The *Pombaline* buildings in Lisbon (named after Pombal) are among the earliest seismically protected constructions in Europe. Small wooden models were built for testing, and earthquakes were simulated by marching troops around them. Pombal also did what was probably the first serious seismological research. He designed a survey that asked everyone in Lisbon whether their animals behaved strangely prior to the earthquake, if the water level in wells changed, how many buildings had been destroyed, and what damage occurred. He is regarded by many as one of the world's first seismologists.

Pombal was nothing if not efficient and effective. Not surprisingly, however, many of his innovative ideas flew in the face of the Jesuits, who saw their power and influence rapidly decline. They had the last laugh over Pombal, however. When King Joseph I died in 1777, his successor, Queen Maria I of Portugal, came to power. She was a devout woman strongly influenced by the Jesuits, and she detested Pombal. One of the first actions she took was to basically ostracize him from all possible political activities. He doesn't seem to have been too bothered, however, and died peacefully in 1782.

Voltaire

Upon hearing the news of the disaster, the renowned French philosopher and author, Voltaire, was struck with horror by the catastrophe and within 10 days had written a long poem about it (*Le Désastre de Lisbonne*), asking how this could be reconciled with a loving God. He argued that the very size of the disaster is "a terrible argument against Optimism," although he went on to add, "The sole consolation is that the Jesuit Inquisitors of Lisbon will have disappeared with the rest . . . for while a few confounded rascals are burning a few fanatics the earth is swallowing up both."

Voltaire's attack on optimism set off a literary and philosophical dispute with Jean-Jacques Rousseau, the leading philosopher of the Age of Reason. Rousseau's response was that physical evils were inevitable, and man only added to them by crowding into cities such as Lisbon. Voltaire's reply was in the form of his famous work, *Candide*. In this ironic masterpiece, Voltaire describes the earthquake and tsunami: "Scarce had they . . . set foot in the city, when they perceived the earth to tremble under their feet, and the sea, swelling and foaming in the harbor, dash in pieces the vessels that were riding at anchor." During the disaster, the only truly good man, honest James the Anabaptist, is killed by the tsunami while saving a wicked sailor, who swims ashore unharmed to try and somehow profit from the disaster. Heavy with irony, Voltaire later describes how to mitigate future seismic disasters, "having been decided by the University of Coimbra that burning a few people alive by a slow fire, and with great ceremony, is an infallible secret to prevent earthquakes."

Indeed, Portugal appeared to be hundreds of years behind the rest of Europe. The Jesuit control of the country was rampant (the Spanish derided their neighbors as *pocos y locos*—few and mad) as the Inquisition searched out not only heretics but also bigamists, witches, Jews, sodomites, and other undesirables.

After the earthquake and tsunami, the people were told that Lisbon had been a very sinful city. The clergy were to blame for permitting outrageous misuse of sacred buildings, and as such this neatly explained why God had destroyed so many churches. It was therefore also understandable that God should not only have destroyed his own churches but also have spared a street full of brothels. God apparently pitied the miserable creatures that frequented such places.

It is indeed true and no less ironic than Voltaire's writings that among the few buildings safely left standing following the disaster were the lightly constructed wooden bordellos of the city. They survived because unlike their opulent, solid-stone-built cousins, they moved in the earthquake and "went with the flow" of the movement as opposed to resisting it and falling down. The brothels were shaken, not stirred.

Most of Lisbon's prostitutes but few of her nuns survived.

This Changed the World . . . of Science

The 1755 Lisbon earthquake represents a turning point for how humans viewed natural disasters. It represents the first attempt to shift away from an "act of God" viewpoint to a more scientific approach. A significant debate raged at the time, as people sought ways to understand how a major European Christian city could be destroyed. Many at the time still viewed the disaster as punishment by God, and this led to a substantial religious revival in Portugal and throughout Europe. Others, such as Immanuel Kant and Rousseau, proposed alternative and more rational views, including describing the disasters as a natural event, and emphasized the need to avoid building in hazardous locations. There was also the need to build properly, a fact to which both Pombal's experiments and the survival of the brothels attest. In many respects, the foundation of modern natural hazards research and the development of our understanding of earthquakes and tsunamis can be traced to this event.

As tsunamis go, the 1755 tsunami was the most significant event to have struck the United Kingdom for millennia. It may not have been anything like that experienced elsewhere, but the United Kingdom is not used to such things. We have to drift well beyond living memory and even written records to come across the United Kingdom's worst-ever tsunami, but this was one that seriously changed the world, at least in northwestern Europe.

13
Storegga
No Referendum for This Brexit

> Plato tells us a story of an island somewhere in the Atlantic Ocean which by a deluge and earthquake was destroyed and swallowed up by the sea . . . which seems to be very applicable to the rupture of the Isthmus . . . that is in the Northern Sea. . . which now meet at the Dogger Sands.
>
> —**Wallis (1701)**

Atlantis?

It all depends on how you read it.

Plato's *Timaeus* has already had a decent workout in this book. In Chapter 11, we duly honored Plato for providing the first written records that ever mention Atlantis around 360 BC. We then jumped into the story of the Santorini eruption, to which the loss of Atlantis seems to be eternally linked.

Well, it is always a fascinating experience to wander down a few literary avenues when searching for useful snippets of information about an event, and this particular avenue produced a wonderful letter written in September 1701 by John Wallis, the Savilian Professor of Geometry at the University of Oxford, to Dr. Hans Sloane, Secretary of the Royal Society, an organization we mentioned in Chapter 12.

Professor Wallis was the most influential English mathematician before the rise of Isaac Newton. He was an interesting character, pushed into the university as a result of the Civil War and the unfolding politics of the day. He did well considering he started off with little mathematical experience. He published some interesting works, such as his 1685 comment in the *Philosophical Transactions* that supported his argument that one's memory is better at night. He reported that he calculated the square root of 3 in his head to 20 decimal places, arriving at the correct answer, 1.73205080756887729353, and retaining it in his mind before writing it down the next day. In short, he had a powerful intellect.

And now back to Plato. Wallis (1701) noted that

> Plato doth there introduce Critias [then an Ancient man] telling a Story, which [when a Boy ten years old] he had heard from his Grandfather [who was ninety years old] of what Solon [long since dead] had told him; namely, than an Egyptian Priest had [long before] told Solon that it did appear from some old Egyptian Records [of which the Greeks had no knowledge] that such a thing had happened in an Age so long before. . . . And all this Tradition [through so many hands, and at such great intervals of time] is, at every step, reported from the Relators present memory. And 'tis very possible, that some one or other of these Relators might so far mistake, or misremember, as to call that a . . . disappearance of an island which was but a tearing of it from the Continent. (Wallis, 1701, pp. 974–975)

In essence, when you read the quote from the beginning of the chapter and couple it with this quote, you realize that back in 1701 Wallis was proposing that the "Chinese Whispers" associated with the long time periods in between each telling of the story may easily have morphed things a bit, as they often do. He therefore was interpreting Plato's "Atlantis" as referring to the destruction of Doggerland (now underwater in the North Sea off England's east coast) by a deluge!

And from our point of view, this is news to us.

The general consensus is that Plato has always apparently been talking about Atlantis and that he was probably really just having a bit of fun. Now here we see that as far back as 1701, at least one British academic was thinking of something much closer to home.

Fascinating.

Dredging up the Past

It is all the more fascinating because in a sense Doggerland *was* destroyed by a deluge. Or at least a tsunami caused significant inundation of this low-lying land at a time of rapid sea level rise. The combined result of these two doubtless saw the accelerated demise of human communities in the area and the veritable collapse of Britain's land bridge with Europe.

This was Brexit on steroids.

As is often the case in our line of business, Wallis' 1701 philosophizing died the death of many great academic ponderings—that of sinking into obscurity and being lost in the morass of old papers, rarely, if ever, to be read again.

A point of note for any budding young earth scientists here—read through the first years of one or more of the great, prestigious journals. These contain papers that were written by geologists who were not really geologists but, rather, polymaths, independently wealthy thinkers philosophizing about things often for the first time and trying to figure out what was going on. They had two distinct advantages back then. First, they had no existing paradigm derailing their thinking. Second, there was more to see then; there was not the vast urban growth we see today. You will find gems of information in these old tomes. Also, if you look at enough of them, you will start to realize that scientists today do seem to spend a lot of time reinventing the wheel. Read the old works first, apart from anything else, it is a good way to breathe new life into some incredibly valuable thinking.

Dogger Bank (this was the generic term for the area for a long time but technically now refers to an upland area of Doggerland) is a large sand bank in a shallow area of the North Sea approximately 62 miles (100 km) off the east coast of England (Figure 13.1-see color plate section). It was well known even back in Wallis' day. Indeed, it had long been known by fishermen to be a productive fishing area and was actually named after the "doggers" or medieval Dutch fishing boats that worked it. However, near the end of the 19th century, there was a growing inkling that this area had once been a fairly extensive landmass. In 1863, Professor King of Queen's University, Ireland, reported on a large quantity of nearshore marine shells that were dredged up from Dogger Bank, and he surmised that the area had once been above the sea and that the shells died when the land became submerged.

Science was moving along.

And then, in September 1931, Skipper Pilgrim E. Lockwood, master of the sailing trawler *Colinda*, L.T. 382, was fishing "halfway between the two North buoys in mid-channel between the Leman and Ower," which was some 25 miles (40 km) off the Norfolk coast, near the Leman and Ower Banks in the southern North Sea. He hauled in his latest trawl and landed a large mass of "moorlog," a local name given to lumps of black wood and peat that they found every now and again. These usually got tossed back overboard because they damaged fishing nets. In this case, before tossing it out, the Skipper decided to break it down in size with a shovel and in doing so he hit something hard that he thought was steel.

This was one of those wonderful moments in science that you wish were recorded for posterity . . . and it was!

When the skipper was interviewed about it several months later, he said, "I bent down and took it below. It lay in the middle of the log which was about

4 feet square by 3 feet deep. I wiped it and saw an object quite black." It was made of antler.

He realized that he had found something unusual, and through a somewhat circuitous route via the ship's owner, the British Museum, and Cambridge biologist Dr. Muir Evans, it ended up in the Castle Museum in Norwich. It was exhibited at a meeting of the Prehistoric Society of East Anglia on February 29, 1932, and attracted considerable attention. The antler bone is approximately 8.5 inches (22 cm) long with a row of barbs running along much of its length—it was a tool made by humans!

It was very quickly termed the "Maglemose harpoon." This discovery was made at a time when radiocarbon dating had not been invented and so the age of the artifact should have been difficult to work out, but as the British Museum noted, it was not archaeologically very unique. Such "harpoons" (pedantically, we now call these a bone point—used singly or tied together to make a fish spear) had been found in both England and northwest Europe and so it could be dated to the Mesolithic (~6,000–12,000 years ago in Britain) and specifically to a culture known as the Maglemose (Danish for "big bog"). Later radiocarbon dating places it roughly approximately 11,700 years old (although there is some debate about this age). This was a time of hunter-gatherers living in small communities with mobile lifestyles who hunted seasonal game, which is rather tough to do when one is underwater.

The Maglemose harpoon, just like Wallis' ponderings, briefly drifted into obscurity as part of a provincial museum exhibit.

However, what made this different was that it was found 25 miles (40 km) offshore. This mystery was not really even addressed by the Cambridge expert Muir Evans, although it was by then recognized that Dogger Bank has been a land surface at one time. He simply stated that "in Neolithic times the Dogger Bank was the northern limit of the land surface which united England to the Continent and a journey from Denmark round what was the North Sea gulf would give these people access to northern England." So, he knew about the existence of a lost landmass, but it was viewed as a route to pass through rather than live in. They were still missing the point.

Once more, obscurity beckoned, and Dogger Bank sunk into an archaeological torpor for decades until eventually in 1998 Professor Bryony Coles, at the University of Exeter, published a synthesis of all the available data on the area and named this lost land "Doggerland" (and now we wait for someone to point out that it was actually H. G. Wells who first mentioned Doggerland in his book *A Story of the Stone Age*, written in the late 19th century).

Archaeological interest surged and geological data started to be incorporated into attempts to map the extent of this landmass. It was now well understood that Mesolithic people had lived, hunted, and died in Doggerland for millennia, and the archaeology was starting to get at some of this detail—complicated greatly by the fact that all of the evidence was underwater. But let us take a detour to geology and the tsunami that has not yet been introduced to Doggerland.

Slip Sliding Away

It all started in the 1980s, so in a sense the geologists were ahead of the archaeologists, who only really started on their side of the equation in the late 1990s. However, neither group knew that both would end up meeting in the middle with two parts of a great story, and so neither one really started first.

In 1985, Professor David Smith and colleagues reported on some finds of an extensive sand layer in northeastern Scotland that they thought might have been deposited by a short-lived storm surge roughly dated to approximately 7,000 years ago. In the 1980s, coastal and marine research was really starting to take off around the North Sea coast, probably because the oil crises of 1973 and 1979 led to a quadrupling and tripling of the world oil price, respectively. North Sea oil was economically viable to get out of the ground, and geology was needed if for no other reason than to better understand the seafloor. It is not surprising that so many of our colleagues are petroleum geologists—when business is good, the funding is amazing!

Studies of the seafloor off Norway led geologists to suggest that three large submarine landslides they had discovered off the continental shelf may well have generated tsunamis. This was soon followed by a seminal paper by Professor Alastair Dawson and colleagues. In many ways, their paper represented the birth of the modern era of the study of prehistoric tsunamis in Europe. They suggested that this proposed storm surge deposit was actually that of a tsunami related to the second of the Storegga (Norwegian for "the Great Edge") slides (Figure 13.2). More evidence, more analyses, and the story started to get fleshed out much more.

Landslides are the second-most important source of tsunamis worldwide—they may generate more tsunamis, but not on the scale of a large subduction zone earthquake that affects an entire ocean, although large ones such as Storegga come very close. We now know that this huge Storegga Slide off the Norwegian west coast took place approximately 8,150 years ago. Also, this was originally called the "second" slide, but recent interpretations show that this was just the most recent of a series of huge submarine landslides that have

Figure 13.2 (a) Position of the Storegga Slide. The arrows indicate the direction of the slide; numbers are wave height in meters. (b) The Storegga tsunami deposit bracketed by peat. Photograph taken at Montrose.
Source: Public domain.

occurred in this area during the past couple of million years or so at a rate of approximately one every 100,000 years.

What is "huge"? Well, this landslide started somewhere around 820-ft (250-m) water depth on the edge of the continental shelf and continued down for thousands of meters. The amount of material that fell is estimated at 840 $miles^3$ (2,400–3,200 km^3)of debris—equivalent to an area the size of Iceland covered to a depth of 112 ft (34 m). Not bad, although it pales into insignificance compared to the largest one known—the 4,800-$mile^3$ (20,000-km^3) Agulhas Slide off the coast of South Africa (this received an honorable mention at the end of Chapter 10). The Storegga is pretty much the next largest.

For a submarine landslide of this magnitude to happen, there normally needs to be some type of preconditioning. For Storegga, this was probably the fact that the sediments that make up that portion of seabed off Norway are composed of interlayered marine and glacial sediments, which respond differently to things such as the amount of weight or "load" on them (added or removed). The marine sediments become relatively more unstable, and if they are on a slope, they can slide.

Various causes have been suggested for this slide. First, the Storegga Slide happened at a time that followed a period when trillions of tons of sediment had been pushed onto the edge of the continental shelf by streams flowing out of melting glaciers at the end of the last glaciation. This alone would overload the sediments. Second, and perhaps in conjunction with the first cause, there may have been an instantaneous trigger such as an earthquake that shook the sediments loose.

Another thought is that an earthquake (again) triggered a catastrophic expansion of methane hydrates, which are created when large amounts of methane are trapped within a crystal structure of water, forming a solid similar to ice. Significant deposits of these have been found in large quantities under ocean floor sediments where conditions seem to be right for their formation. For the record, a cubic meter of solid methane hydrate expands to approximately 164 cubic meters of methane, so such an event would catastrophically destabilize undersea sediments. Sometimes these solid deposits cap even larger deposits or "bubbles" of methane gas. Methane hydrate release features are often reported from studies of the ocean floor, normally close to the edge of the continental shelf. These are a bit like bomb craters and form when the cap collapses and a large "bubble" is released; bubble releases alone may generate tsunamis, and they have been suggested as the cause for several ancient tsunamis throughout the world. Bubbles and a landslide in this instance may have worked together to generate a really big event.

Whatever the actual trigger (this is discussed later in the chapter), this was a huge event. Because the landslide occurred underwater, the sediments, once they started moving, kept on going for approximately 250 miles (400 km). Traces of the tsunami caused by this huge landslide have been recorded in Scotland up to 50 miles (80 km) inland, but sediments deposited by the waves have also been identified along most of the coastlines of the North Sea, including the Shetland Islands, Faroe Islands, Norway, England, Denmark, and even across the Norwegian Sea to Iceland and Greenland. There are undoubtedly still many places yet to be discovered. For example, little attention has been paid to the potential impact of the Storegga tsunami on Ireland. Recent computer models suggest that waves approximately 16 ft (5 m) high would have struck the west coast of Ireland and doubtless penetrated into the Irish Sea between Scotland/England and Ireland. The problem is that it is difficult to find evidence for the tsunami because since the Mesolithic time when this catastrophe happened, coastlines in this area of the world have pretty much all been drowned as sea level has risen by up to 65 ft (20 m) in some places. It is difficult to find geological evidence for a wave 16 ft high when this currently could be 50 ft (15 m) underwater. However, there is possible evidence in the Shannon estuary in southwest Ireland; if this evidence is real, then much of Ireland would have been affected by the waves.

Given that there have been regular submarine landslides in the area, albeit with quite a long time between them, the worst thing we could do it upset the environment there. So, sure enough, that's what we have done.

Hey! We found natural gas, a.k.a. methane! No surprise there. And so the Ormen Lange gas field came into being in September 2007. It is located *within* the slide scar of the Storegga Slide where seabed depths vary between 2,600 and 3,600 ft (800 and 1,100 m). Wonderful news though—this would supply Great Britain with enough natural gas to cover 20% of its total annual needs. Phew, that's comforting.

Perhaps the better news is that a thorough geological survey concluded that the Storegga Slide was caused by sediment that built up during the previous glacial period and its rapid demise, and so such an event is unlikely to happen again until after a new ice age. Thus, it was decided that the Ormen Lange gas field would not significantly increase the risk of triggering a new slide.

As an interesting aside, the gas field is named after Olaf Tryggvason's longship, the Ormen Lange. Olaf was a 10th-century Viking king of Norway who was either born in the Orkney Islands north of the Scottish mainland or arrived there when he was 3 years old. A staunch Christian, in later life he baptized the explorer Leif Ericson, who took a priest with him back to Greenland. Olaf had a great influence throughout the North Sea region and was brutal in torturing those who refused to convert. In one instance, he used a wooden pin to pry open a victim's mouth and inserted a snake through a drinking horn; the snake slithered into the victim's mouth and down his throat. The victim died.

The parallels with the influence of the Storegga Slide are quite intriguing. It too had a widespread influence on communities throughout the North Sea and as far away as Greenland, and it certainly killed many people. The gas field seems aptly named.

Back to the Storegga.

Doggerland and Storegga Meet

As tsunamis go, this was a pretty big one, but it also happened at a time and a place when humans would have been particularly exposed. The timing meant that in the early Holocene (which began ~10,000 years ago) this area of northwest Europe was a feeding ground for Mesolithic human populations following the retreat of the glaciers. Many lived on the low-lying ground that surrounded the North Sea—which was considerably smaller than it is today but was slowly growing southwards as the sea level rose.

In Britain, the Mesolithic dates to between approximately 4,000 and 10,000 BC, and it represents the period immediately before the introduction of agriculture. These were small hunter–gatherer communities whose mobile lifestyles allowed them to make use of the wide range of seasonal game in northwest Europe. But it was a difficult time to be living in this area; times they were a-changing.

Research has pointed to two possible impacts on human populations at this time in and around Doggerland, and this brings in a new wrinkle we have vaguely mentioned already together with our archnemesis, the Storegga Slide.

Sea level—the new wrinkle—was rising rapidly and Doggerland was getting wetter. As with all of such things, when scientists say something such as "sea level was rising rapidly," we generally default to what is going on today. Climate change-driven sea levels are rising rapidly today, with predictions of a rise of up to 4–14 ft (1.4–4.3 m) by 2300 for London, equivalent to approximately 0.17 inches (14.3 mm) per year. Almost imperceptible to us but with devastating effects over two or three generations. We know this, but because it is so slow, even if it is geologically rapid, it is difficult to get our heads around it sometimes.

One obvious thing that comes out of this is that clearly sea level rise does not occur smoothly. Today it is accelerating. Back in Mesolithic times, the same problem was occurring, but it was a *major* problem. Approximately 8,200 years ago in the area of the Northern Hemisphere we are talking about, the rise in sea level could have been up to approximately 21 ft (6.5 m) in less than 140 years—about 2 inches (46 mm) per year. Now that is fast!

This incredibly rapid rise in sea level also happened at a time of rapid cooling. Definitely not a good time to be around Doggerland or anywhere else in that neck of the woods. It has been speculated that this double-whammy could well have been caused by a large pulse of meltwater from the final collapse of a large ice sheet in northeastern North America. For thousands of years, this ice sheet—the Laurentide Ice Sheet—covered a big chunk of North America. It did a great job of blocking rivers, but as it started to melt, huge lakes formed at the margin of the ice sheet because the rivers had nowhere to go; they were still blocked. Glacial Lake Agassiz was the largest of these, more than twice the size of the largest lake in the world today. Occasionally as the ice was melting, the water in this lake would find a way out, but the last and largest of these outbursts released a massive amount of water. More than 36,000 miles^3 (150,000 km^3) of water flowed through Hudson Bay into the Labrador Sea in only 6–12 months.

Such a massive volume of water being released in a short period of time did several things. First, because this book is about tsunamis, it is interesting to note that when such a massive volume of water is so rapidly pumped into a coastal sea such as the Labrador Sea (between Greenland and northeastern Canada), it has the potential to generate a tsunami. Recent research suggests that it could have been at least 6 ft (2 m) high and traveled some 30–60 miles (50–100 km) along the coast—not really one that would change the world, but an interesting side effect of this catastrophic outburst of water.

In addition, and definitely much more important, this huge pulse of meltwater may well have affected the North Atlantic thermohaline circulation, reducing the Gulf Stream's northward heat transport and causing cooling around our area of interest of between 1.8° and 9.0°F (1° and 5°C).

Back in Doggerland, the impact of this abrupt cold event, rapid sea level rise, and the Storegga Slide tsunami on the Mesolithic people was devastating. As sea levels rose and the North Sea slowly flooded southwards, Doggerland eventually became just the island of Dogger Bank. This had definitely happened by 8,000 years ago, and by approximately 5,000 years ago even that was lost to the waves. The extent of the impact of the Storegga tsunami is still something of a mystery, although largely because just as in the case of Ireland, all the evidence is underwater. It is also a problem to try to model what happened because sea levels were changing so rapidly that it is difficult to know where the coast was at any one time. However, computer models have been produced, and under one of several scenarios, tsunami waves up to 29 ft (9 m) high strike the northern islands of the Doggerland archipelago, completely inundating some of them and propagating up to 13 miles (21 km) inland. Up to approximately 770 miles^2 (2,000 km^2) was flooded (35% of the land surface). None of the scenarios show a catastrophic inundation of the entire island archipelago. This sounds a bit grim, but perhaps it was not so bad in the sense that Doggerland was very low-lying anyway, so a decent king tide or a large storm surge may well have had a similar effect . . . possibly. But the tsunami was probably a bit worse.

Recent research has revealed that at the head of one of the old drowned river systems offshore from The Wash in eastern England, there are remnants of Storegga tsunami sediments surviving. The evidence was found in a series of sediment cores that allowed researchers to examine the environmental after-effects of tsunami inundation preserved in the pollen record. The effects were temporary and localized because not all of Doggerland was inundated. Storegga undoubtedly destroyed large areas of forest, opening up the landscape, but because a terrestrial signal remains after the event, it was not a case of the tsunami wiping the slate clean. This eventual honor would go to sea level rise.

However, the flooding of Doggerland by Storegga would have marked a significant moment in the demise of Mesolithic use of the area. The loss of habitat, lives, communities, and most likely many communication routes would mark the first stages in the creation of the North Sea. Whenever it became official, the North Sea would have played a key role in the changing cultures of the Atlantic region of northwest Europe. In the early days there was the development of a barrier to contacts maintained over generations both across and within Doggerland. Later, however, the North Sea becomes the means of communication—by boat. In this period of transition from dry land to sea, there may also have been a transition from movement by land to movement by water. It had to happen sometime, and the oldest recovered boat in the world, the Pesse canoe, was found in the Netherlands, not far from Doggerland. It is a dugout canoe that was constructed around 9,500–10,000 years ago, and so the technology was there for this to happen.

This process of change would have obviously been a gradual one, but it seems likely that the Storegga tsunami would have had a similar effect to that of the Santorini tsunami on Crete (see Chapter 11). It provided a devastating one-off blow to Mesolithic communities living on a slowly flooding, low-lying bit of land. It did not cause the end, but it undoubtedly changed the environment for them, quite possibly causing the initial severing of ties between Britain and Europe by eroding the land and creating new waterways. After this, it was one-way traffic, no reunification, no bridges. The umbilical cord to mainland Europe was gone, and the Mesolithic Brexit was complete.

Afterword

The 8,150-year-old Storegga tsunami is not alone. There have been more recent ones caused by submarine landslides, but they have been smaller . . . so far.

And so we have submarine landslides that can cause devastating tsunamis in the most unexpected areas of the world. There is perhaps nowhere better in the world to find these than in the Pacific Ocean. It is full of them. There are well over 20,000 islands in the Pacific, and most of these are volcanic in origin. This means that they have grown up from the seafloor and reached the surface; it also means that they can get pretty large because the Pacific Ocean is deep. For example, Mauna Loa on the Big Island of Hawai'i is higher than Mt. Everest; it's just that most of it is underwater. And just like any mountain, bits have a habit of falling off the sides from time to time. For example, the Alika Slide happened approximately 120,000 years ago. It was *only* a 360–480 Cubic Mile (1,500–2,000-km^3) submarine landslide, but it is thought to have caused a giant tsunami that washed up 1,066 ft (325 m) high on the island of Lanai some 100 miles to the northwest. Not bad. Now, multiply that up by 20,000 islands and you start to see the scale of the problem for the Pacific. And the poor Hawaiian Islands are stuck right in the middle. Not only are they doing a great job of generating their own tsunamis but also the Pacific Ring of Fire surrounds them on all sides. These islands have the potential to be struck by tsunamis from all points of the compass, but nothing quite prepared them for what came "out of left field" or, in the case of the 1960 tsunami, the back right.

Read on.

14

1960 Chile

Did the Earth Move for You?

> The original design-basis tsunami for Fukushima Daiichi of 3.1 meters was chosen because a 1960 earthquake off the coast of Chile created a tsunami of that height on the Fukushima coast. However, greater attention should have been paid to evidence from further back in history. Over the last decade or so, evidence of much larger tsunamis in and around Miyagi has emerged. Japanese researchers have discovered layers of sediment that appear to have been deposited by tsunamis and have concluded that the region had been inundated by massive tsunamis about once every one thousand years.
>
> **—Acton and Hibbs (2012, pp. 11–12)**

Hubris?

One of the most common statements that tsunami researchers come across either from working in areas struck by a recent tsunami or in trying to extract funding from assorted agencies in an attempt to examine evidence for past events is something like the following: "I have lived here all my life and never seen anything like this." Whether such thinking leads to hubris or simply a blinkered approach to the whole tsunami hazardscape is uncertain, but what it does inevitably lead to is ignorance, inaction, and, unfortunately, in many cases preventable deaths on a massive scale.

When most people think about tsunamis, they think about Japan, which until the 2011 Tohoku-oki tsunami was considered to be the most tsunami-prepared country in the world. Indeed, Japan is and was prepared, but the 2011 caught the country by surprise. The cause for this surprise is deeply rooted in hubris. In 1952, Japan established a tsunami warning system to mitigate damage caused by tsunamis associated with large earthquakes occurring close to its shores. This was years ahead of most other Pacific countries. However, this was of little use in 1960 when an unexpected tsunami struck the country, killing at least 139 people and destroying well over 1,500 houses. There were two very significant outcomes from this disaster. First, Japan realized that there needed to be international cooperation in order for it to be prepared for tsunamis that came from beyond its shores. Second, the magnitude of waves associated with this surprisingly large event came to be the design basis for the construction of nuclear power plants such as Fukushima Daiich. This had devastating consequences during the 2011 event.

One of the problems here is that the 2011 Tohoku-oki tsunami was exactly what Japan had originally set up its tsunami warning for—a locally generated event. But many of the assumptions about how big tsunami waves might be along the coast were based on events such as the unexpected 1960 tsunami, a distantly generated event that is discussed later. What Japan should have been aware of is that well before the 2011 tsunami, Japanese researchers had discovered and reported geological evidence for massive tsunamis far larger than the 1960 one inundating the coastline at least once every 1,000 years. Locally generated events, at least in Japan, can be far larger than their distantly generated counterparts.

Another problem is that the lack of an international tsunami warning network meant that distant tsunamis could be devastating because there was no forewarning available for coastal residents and so a massive earthquake that occurred 10,500 miles (17,000 km) away from Japan generated a tsunami that killed well over a hundred Japanese people. Fortunately, we now have a far better tsunami warning system, but what of the 1960 tsunami—just how bad was it?

1960 Chilean Earthquake and Tsunami: The Earth Moved

On May 22, 1960, a giant magnitude 9.5 earthquake, the largest earthquake ever instrumentally recorded, occurred off the coast of southern Chile. There had been some large foreshocks that unfortunately were not known to be warning of a bigger one to come. The largest of these was a magnitude 7.5 that occurred on May 21 and saw approximately 93 miles (150 km) of seafloor rupture, but there was no tsunami. However, this paled into insignificance to the one that would happen the next day that ruptured nearly 620 miles (1,000 km) of seafloor, generated a tsunami well over 50 ft (15 m) high in Chile, and moved the entire country to the west in a matter of minutes as the ground shook.

This earthquake generated a tsunami that was destructive not only along the coasts of Chile and distant Japan but also across the entire Pacific region, including the Hawaiian Islands in the middle, New Zealand and Australia to the southwest, California in the northeast Pacific, the Philippines in the west, Alaska in the far north, and many Pacific Islands in between (Figure 14.1-see color plate section).

Estimated casualties for Chile range between 1,655 and 6,000 deaths, with $2.9–$5.8 billion in damages and more than 2 million people left homeless; these figures are based on a combination of the effects related to both the earthquake and the tsunami. Beyond Chile, such figures are even more difficult to estimate but are approximately 230 deaths and $950 million in damages, and these relate only to the tsunami. Monetary estimates are based on 2011 values. In Hawaii, the tsunami caused 61 deaths and more than $23 million in damage. A further $1 million in damages and 2 deaths occurred along the west coast of the United States, and 21 people died in the Philippines.

An Exceptional Event?

A subgroup of the Chile's indigenous Mapuche people lived on the coast adjacent to the earthquake's epicenter. They were called the Lafkenche, which means "people of the coast."

The earthquake and the tsunami were not unexpected to them. Their Machi or shaman/wizard dreamed about a large earthquake and predicted it would happen. But this is simply a single part of a series of complex indigenous knowledge and beliefs. There is a tradition that these major events happen in a cyclical manner approximately once every 100 years. Scientific research in recent years has revealed that previous large earthquakes (and tsunamis) occurred in the region in 1837, 1737, 1657, and 1575, or if you wish, about once every 100 years—we got there in the end. This confluence between indigenous knowledge and scientific research is a theme that we have discussed previously in this book. The depth of knowledge held by indigenous peoples is only starting to be realized, with Western science slowly recognizing that there is much to learn from indigenous knowledge.

Perhaps not surprisingly given the magnitude of the 1960 earthquake and tsunami and predecessors, such events form part of the Mapuche creation tradition. This concerns a fight between two snakes: Cai-Cai and Treng-Treng. Cai-Cai is a snake that comes from the sea. It is a bad snake. Treng-Treng comes from the highlands and is a good snake. The Mapuche creation tradition is represented by the fight between good and bad, between the earth and the sea (water). During the fight, the first humans ran uphill to escape the sea and avoid drowning. The hills rose as the water rose, and they ran uphill so far that they almost touched the sun. Only a few families survived, and they were the first of the new cycle. This new cycle began at the end of the old, between the shaking and moving of the earth and the rising and falling of the seas. This tradition resurfaced within the Mapuche following the 1960 earthquake and tsunami.

Intriguingly, in Mapuche tradition such earthquakes and tsunamis were also associated with the will of Pillán, a spirit who inhabited volcanoes. All volcanic activity is related to the will of that spirit. So once again, the Mapuche were not surprised when 2 days after the 1960 earthquake and tsunami, Puyehue volcano erupted.

Beyond the shores of mainland Chile lie its offshore islands—Easter Island (Rapa Nui) in particular. Although it is Chilean, Rapa Nui lies 2,100 miles (3,500 km) west of the mainland. It is a little like the Hawaiian Islands in that it sits in the Pacific Ocean exposed to tsunamis coming from almost anywhere in the region. In just the past century, Rapa Nui has been struck by at least the 1946 tsunami all the way from Alaska and the 1960, 1995, 2010, and 2015 Chilean events. The 1960 tsunami took approximately 6 hours to reach the island with waves at least 20 ft (6 m) high inundating the coast. It was an unexpected and traumatic event.

We know from archaeological research that the vast majority of the prehistoric Polynesian community of the island lived in coastal areas, with the inland mostly used for agricultural purposes. Most of the island's population was therefore exposed to the possibly devastating effects of prehistoric tsunamis. This may help explain some of the turmoil that occurred there in the past.

From geological and historical research, we know that probably up to as many as 50 tsunamis have affected the island during the past 1,000 years. This includes the major precursor event to the 1960 Chile tsunami that occurred in 1575. It also includes the large prehistoric tsunami discussed in Chapter 4 that affected numerous Polynesian communities in the 15th century.

There is considerable debate regarding the human settlement of Easter Island. There are three main theories. One line of research proposes that the first settlers arrived between 800 and 1200 AD from East Polynesia. Rapid deforestation took place, resulting in an ecological, and consequently cultural, catastrophe. Another suggests a later arrival between 1200 and 1300 AD, with the rest of the story pretty much the same. The third finds early signs of human activity around 450 BC. This much earlier date derives from the discovery of an American human-dispersed weed in lake sediments on the island. The debate continues, but although all seem to agree that major landscape changes took place, they can't agree on the timing. There is even debate regarding whether it was a slow or rapid change, but change it did.

Looking beyond the shores of Easter Island and putting changes in Polynesian culture and society in a wider context has allowed researchers to better see "the forest from the trees." It was in the early 19th century that most of the first written descriptions of the inhabitants of Pacific islands were made. Here, one of the overarching observations was the surprise that most settlements were inland and also fortified. Conflict was widespread. This 19th-century status quo marks a profound prehistoric societal transformation or collapse that appears to have occurred in prehistory throughout the South Pacific from the Solomon Islands in the west to Easter Island in the east and from the Marshall Islands in the north to New Zealand in the south. In Chapter 3, this region-wide cultural change was lined up directly with a major 15th-century earthquake and tsunami along the Tonga Trench. In the isolated Polynesian outlier of Easter Island, there is archaeological evidence that the manufacture of spearheads (*mata'a*) began at approximately this time, confirming that destructive intertribal warfare had begun.

The existing archaeological debate as to what happened to Easter Island's prehistoric Polynesian culture still fails to consider the impact of tsunamis. And yet here we are. Both geological and archaeological evidence exist for a region-wide catastrophe throughout Polynesia in the 15th century. Furthermore, historical evidence exists for the 1960 tsunami devastating the island's eastern coastline, so much so that it destroyed the famous Ahu Tongariki (Figure 14.2). It was not until a few days after the tsunami that the devastation was seen, with estimates of waves exceeding 32 ft (10 m) in height with a run-up in the order of 1,600–2,000 ft (500–600 m) inland toward the base of the Rano Raraku volcano. The *Ahu*, which is the mound or stone pedestal on which the Moai rest, was completely destroyed; several Moai weighing up to 50 tons were carried up to 300+ ft (100 m) inland; and the whole area was littered with tsunami boulders, dead sheep, ripped-up seaweed, fish, and human bones from the tombs that had been under the Ahu.

It has since been restored and is a major tourist attraction, so if you have visited the island, this will give you a good sense of the power of the 1960 tsunami. We can prove all of this to those who do not believe it because there are, very fortunately, before and after photographs taken at the time.

However, what about earlier events? What would an earlier event of similar or greater magnitude have done to the then coastal population and resources?

The evidence that prehistoric tsunamis may have left their mark on Easter Island is slowly coming to light. Several Ahu have been found to be composed

Figure 14.2 Ahu Tongariki, Easter Island. (a) Ahu Tongariki panorama looking northwest, 1914–1915. (b) Pre-1960 tsunami photograph taken looking northwest from Ahu showing pre-existing collapsed Moai. (c) Post-1960 tsunami photograph taken from inland plain showing moved and broken Moai. White ovals in b and c mark similar background scenery visible in both photographs.

Sources: (a) British Museum, public domain; (b and c) photographs taken by the Chilean artist Lorenzo Domínguez—permission for use granted by The Estate of Lorenzo Domínguez, holders of the copyright.

of ancient recycled Ahu remains and marine material. Given the historically documented evidence for the effects of the 1960 tsunami, it has led some archaeologists to propose that all of this material was most likely moved inland by one or more past tsunamis and then recycled during the reconstruction process.

The world of tsunami research is replete with received wisdom. For example, it is often argued that tsunamis cannot have caused widespread damage and cultural and settlement pattern changes because alternative archaeological narratives for such evidence already exist. However, narratives aside, the reconstruction of the histories of ancient Polynesian societies such as that of Easter Island is extraordinarily complex. There is a lack of written records and archaeological evidence is fragmentary across a wide region, but all existing archaeological narratives were created before scientists had the ability to identify the geological records of past tsunamis. The debate over the fate of the Polynesian population of Easter Island therefore focuses mostly on the three narratives mentioned previously. None of these has incorporated the possible impact of tsunamis on the island.

An interesting aside is that Ahu and their associated Moai are considered to be sacred ceremonial sites. Now if this was an episode of that wonderful TV archaeology program *Time Team America*, the viewer might not be surprised to hear one of the archaeologists suggest that given the incorporation of marine and reworked Moai material, such reconstructed Ahu sites might be viewed as ritualistic sites honoring the sea. Could it be that simple? Could the damage caused by the 1960 tsunami be pointing toward a survival strategy steeped in Polynesian cultural practices?

Traditions and mysteries of indigenous and prehistoric communities in Chile put the effects of the 1960 tsunami in a deeper time and deeper cultural context. But what about those people living on shores well beyond Chile?

The Hawaiian Islands are even more exposed to tsunamis than Easter Island. Sitting almost equidistant from several major sources along the northern margins of the Pacific Ring of Fire, we have already discussed the effects of the 1946 event that came from Alaska more than 2,400 miles (~3,900 km) away from them. Now here we have one from the southeast, a mere 6,700 miles (10,800 km) away—almost three times farther. Surely, the tsunami could not have been anywhere near as bad?

The Shinmachi Context

Japanese laborers arrived on the Big Island in 1885 to work in the sugar cane and pineapple plantations as part of the Kanyaku Imin immigration system. This system represented an agreement between the Japanese government and the Kingdom of Hawaii. Being foreigners in a foreign land, they tended to stick together and so by about 1900 they had set up their own community, Shinmachi ("New Town"), on Hilo's low-lying waterfront land on the east coast of the island. Forty-six years later, this thriving community was obliterated by the 1946 tsunami that came from the north. Almost immediately after the event, in an act of solidarity and community spirit, they rebuilt on the same vulnerable, low-lying lands. Shinmachi rose from the chaotic debris of the tsunami only to be struck down for a second and final time by the 1960 Chilean event. It was never rebuilt, and the area was turned into the Wailoa River State Park, a complex of sport fields, picnic areas, and parking lots. This became part of Hilo's tsunami buffer zone, an area given over to future tsunamis so that they can expend their energy inundating areas where no one lives. People on the Big Island learned a lot that day.

At 11:30 a.m. on May 22, 1960 (Hawaii local time), the authorities knew that a large earthquake had occurred in Chile, and a tsunami watch was issued with a predicted arrival time approximately 15 hours later, around midnight. The tsunami consisted of three main waves. The first—a small one—arrived just after midnight. The second one was twice as high, around 9 ft (2.7 m), and it flooded Kamehameha Avenue and the business district of Hilo. The third and largest wave arrived about 1 hour after the first as a 20-ft (6.1-m)-high nearly vertical front. It rose to a height of 35 ft (10.7 m) and moved boulders of up to 22 tons from the offshore breakwater into downtown Hilo. By 2:15 a.m., the wave heights of all the following waves had reduced to the point that people could re-enter Hilo. Unfortunately, despite the first evacuation siren sounding at 8:35 p.m., 61 people lost their lives and property damage was estimated at $23.5 million (1960 value). What went wrong?

Did Hawaii Learn from the 1946 Tsunami?

It was the 1946 tsunami from Alaska that really put tsunamis on the map for the Hawaiian Islands. It was largely responsible for the development of the tsunami warning system, modern scientific studies, and increased public awareness. This does not mean that they all happened immediately, and in reality they are always going to be something of a work in progress.

The 1960 tsunami was the next major event to strike the Hawaiian Islands. There had also been a smaller tsunami in 1957, but the 1960 one was the next big test—a test that saw the demise of Shinmachi and much, much more (Figure 14.3).

The Hawaii County Police Department had been put on alert early, and it, together with the Hawaii Civil Defense Agency, thought that the best way to warn of the impending arrival of the tsunami was to send an officer to South Point, the most southerly point of the island, to watch for it. The reasoning was that they thought South Point was closer to the source of the tsunami that was coming from Chile, not realizing just how far east (as well as south) South America is from Hawaii. Because there was no telephone at South Point, the officer was instructed to send up a flare when the waves started to come ashore. This was not a great plan given the almost endless list of things that could go wrong, but by 10:45 p.m. the authorities had a better idea and had found an amateur radio operator near South Point who could send messages to both Hilo and Honolulu on the main island of Oahu, approximately 220 miles (350 km) to the northwest. The authorities had no notion that by being

Figure 14.3 (a) The island of Hawaii. (b) Looking west toward downtown Hilo after the May 23, 1960, tsunami.

Sources: (a) J. Goff; (b) courtesy of the Pacific Tsunami Museum, photograph number 1960.04, Polhemus Collection.

on the eastern side of the island the main population center of Hilo was closer to Chile and would be struck first before the deserted coastline of South Point.

Midnight arrived and nothing had happened except for reports of small waves arriving at some of the other islands, and so officers such as Bob "Steamy" Chow thought that this was just another false alarm. His detail had been to evacuate downtown Hilo, but when everyone thought the danger was over, people started to return to their homes, and there was simply insufficient manpower to stop them.

The first two waves had already struck Hilo just after midnight and didn't cause much damage, and at 12:52 a.m. a report came in from South Point advising that there was "No Activity." This was soon to prove to be far from the truth for Hilo. The third wave was about to strike.

Saved by a Light Pole

Mark Olds lived very near Hilo Bay. He was one of the many residents who evacuated the danger zone, only to return before the danger had passed. He had finally left his home after hearing the second warning broadcast over the radio, but then he went to his office in downtown Hilo—not exactly a safe place. He continued listening to the radio and he too heard that only small waves had struck Tahiti more than 2,500 miles (4,000 km) to the south. Surely it must have been bigger there, it was closer to Chile? Thinking that this was just another false alarm, he returned to his house by the bay. Around midnight, he was ready to go to bed but on impulse decided to take a last look at the ocean before retiring. Opening his back door, he switched on a light, revealing that the yard was completely flooded.

It was time to leave.

As he reached the *lanai* (verandah), the nearby electricity plant exploded. He was now rather highly motivated and raced through the darkness to his car, where his keys were hidden behind the visor. But in the dark his hand hit the visor and the keys fell to the floor. The next few minutes seemed like hours as he searched the car floor with trembling fingers. Finally, he found the keys and started his car, but just at that moment another car rushed in and parked across his driveway, blocking his exit. The driver rushed across the street and began to climb a utility pole shouting, "The wave is coming now. Get up a pole!" Obediently, Mark began to climb the pole. "Not this one, not this one, go to the other pole!" Too bemused to argue, Mark jumped to the ground, ran to the next pole, and scampered up it just ahead of the rising water. He climbed approximately three-fourths of the way to the top as the water rose to his feet. When it had subsided, Mark and his new acquaintance left their perches and drove off in their respective cars.

This was not a good plan.

The water rose again, causing Mark's car to stall. He then heard calls for help from nearby, and he left his car and went to a house to assist a woman who was trying to escape from her upstairs window. The woman jumped from the window, landing on top of Mark, immersing them both in the muddy flow. Somehow after this comedy of errors, they both managed to escape the water and make it to higher ground and safety.

Someone who probably should have known better, however, was Everett Spencer, who lived with his family in downtown Hilo. Everett had been a student at Laupahoehoe High School in 1946 and had run from the tsunami across the playing field and seen the devastation that it wrought. Now, despite the tsunami warning, he did not expect the predicted wave to have a serious effect where he currently lived and so, feeling relaxed about the whole thing, he went to sleep as usual—it was late and hubris had set in. He woke with a jolt to find his roof gone and a bright blue light in the sky, and he could hear a deep rumbling noise as his house was actually moving. When the house finally grounded again, Everett called for his children; all were safe except for his 1-year-old son, who couldn't be found. Once he got his family safely to high ground, he ran back to look for the infant and was eventually rewarded for his endeavors with the sound of a baby crying from the direction of a neighbor's house. There, he found his son and pulled him from the debris, bleeding. He ran with his son to the nearby home of Dr. Francis Wong, who was still sleeping, completely unaware of the tsunami. When they all finally arrived at Hilo Memorial Hospital, they were the very first patients— no one there knew about the tsunami!

Mrs. Fusayo Ito: Watched over by Buddha

During the 1946 tsunami, many residents of Waiakea (to the east of downtown Hilo) had run to the base of a large mango tree on high ground near Mrs. Fusayo Ito's home, where they had been safe. Not surprisingly, for that reason, Fusayo had decided to stay in her "safe" house, despite her daughter's plea to move to her home high above Hilo Bay. That night, Fusayo heard neighbors walking past her house heading to the Wailoa River to watch the predicted tsunami. She would have liked to join them but didn't want to venture out in the dark, and so she watched from her door until approximately 12:30 a.m., when she heard the watchers saying, "Oh, we not having tidal wave today, all finished."

Half an hour later, she heard the sound of an explosion, "like a bomb," and everything went dark.

In the next instant, the third wave entered her open door, seized her, spinning her around and around. She was hit on the head, fell through a hole, struggled to lift herself, and lost consciousness. She said, "And then when I woke up. I was surrounding with the debris. . . . I cannot see the water, just all covered with the wreckage." She moved her legs trying to find a foothold, but there was only water underneath her. She began to float on her back but then was deluged by another wave, swallowed more water, and smelled gasoline. Now floating on a piece of debris, she heard whistling nearby and a man's voice called out "Can you swim?" "No," she replied. "Then hang on!" She held on for dear life and found herself being dragged between the tops of two large pine trees. "I was washed down. And went pass the park and out in the ocean."

As Fusayo was being washed up the coast, she could see the headlights of cars moving on the road above the sea cliff. As the debris began to spread out, she realized that all she was floating on was the frame of a window screen. Only a tiny woman like herself, a mere 4 ft. 11 in. (1.5 m) tall, could have been kept afloat by such a flimsy structure. As she continued moving up the coastline, she also moved out to sea:

> I was washed way far, so many miles out. I cannot see the lighthouse anymore. Just the sky and ocean, sky and ocean. And I prayed "My Lord Buddha," and I wasn't even afraid of the ocean. The ocean was beautiful. I look in the sky everything beautiful.

Fusayo accepted that death would come eventually, by sharks or by drowning, but felt no concern. She had no control over whatever might happen.

Monday, May 23 dawned in Hilo and Fusayo was still alive, still floating offshore. The morning tide had carried her back toward Hilo. As the morning light spread across the water, she saw that most of the debris had dispersed during the night and that she was alone on her window screen. Eventually, she saw something white and wondered if it could be a "ghost ship," a figment of her imagination. In fact, it was the Coast Guard's 95-ft patrol boat under the command of Chief Boatswain Fredrick R. Nickerson. She was lifted aboard, given first aid, and wrapped in a blanket. When Fusayo had heard and seen her rescuers, her peace and resignation instantly left her, "I just cried, cried, cried. Cause I thought I'll never will see a human being again, because I never will come back to this world." Despite her long ordeal, her only physical injuries were a cut finger and bumped knee, but the effects of the shock were great and for many weeks merely the sound of water would set her shaking.

Following the tsunami, there was a pleasant surprise for Mrs. Ito. Like many, she had lost her home, but amazingly all of her important papers survived, including her bank bonds. Perhaps because of her experiences in 1946, all of these had been kept in a waterproof bag. A bulldozer driver involved in the clear-up work near her house found it and turned it in to the police. It was a wonderful surprise when the police called Mrs. Ito, "I cried and cried—if I was dead, I don't need those things, but if I am alive, I need them." Indeed, like many others trying to reconstruct their lives, she had earlier tried to claim the value of her bonds from the bank but had no serial numbers. Once found, it took her days to soak the mud from the bonds, but in the end the serial numbers were visible and the bonds were honored by the bank. Not only was Buddha looking after her, but her wise move to keep her valuable papers in a waterproof bag meant that she was prepared for just such an eventuality, even if she never quite expected it to be so dramatic.

It is perhaps extremely fortunate that there are so many survivor stories from Hilo. There are many more and many poignant stories that cover almost every aspect of the human tragedy that unfolded that day. Why the focus on the Hawaiian Islands? Why are there so few stories from elsewhere? This is almost entirely due to the Pacific Tsunami Museum that was co-founded by Dr. Walter Dudley and Jeanne Branch Johnston. In 1988, Dr. Dudley published a book, *Tsunami!* in which he made a request to the Hawaii community for survivor stories. In 1993, Branch Johnston, a tsunami survivor, saw the need for a tsunami museum in Hilo. The museum opened its doors to the public in June 1998. The museum continually expands its exhibits and content to capture both the human stories and scientific explanations of the tsunamis portrayed, in keeping with the museum's mission to educate the public and save lives in the event of a tsunami. These survivor stories now extend to more recent global events as well and are also discussed later in this book.

The Aftermath

The 1960 tsunami had given scientists another "live experiment." Immediately following the tsunami, researchers set about collecting data. The results of the research show a striking contrast between the 1960 Chilean and the 1946 Alaskan tsunamis. The 1946 waves reached a maximum of 55 ft (17 m) and averaged 30 ft (9 m) along the northeast coast of Hawaii. This is the coast facing directly toward the source, and the heights on this side of the island averaged more than twice that of the island as a whole. A similar increase in wave height might have been expected along the southeast coast in 1960; this is the side of the island that most directly faces Chile. Yet waves were no larger there than along the more protected west and northeast coasts. Specialists believe that these variations in wave pattern might be caused by the difference in cross section of the Hawaiian ridge (the long undersea mountain chain from which the islands poke up above the surface of the sea) as encountered by waves approaching from different directions. The ridge presents a barrier of almost continental dimensions to Alaskan tsunamis, which approach it broadside, but only a small barrier to Chilean ones, which approach it end-on. Another factor that might have led to larger waves along the northeast coast of Hawaii is the relatively shallow, sloping shelf that extends outward from the sea-cliffed Hamakua Coast just north of Hilo and is absent elsewhere on the island.

Based on their studies, scientists concluded that although each tsunami is indeed unique, the location of a tsunami's source may be one of the major factors in determining which area will be most severely affected by the waves and how great the subsequent damage will be. Tsunamis from near the same geographical place of origin tend to produce remarkably similar relative patterns of wave height at any one location. For example, tsunamis affecting Hilo in 1946 and 1957 both originated near Alaska, and although the maximum wave heights in Hilo from the two tsunamis were very different, the same areas of the city were most severely affected by both tsunamis. The 1960 Chilean tsunami, on the other hand, produced a very different pattern of inundation from those of either of the earlier tsunamis. In other words, the patterns of wave heights and hence flooding on Hawaii's shores produced by tsunamis of different geographic origin are strikingly different. Tsunamis from nearly the same origin, although perhaps differing in relative severity from place to place, tend to be more similar than previously thought.

An interesting point here is the question, Where is the evidence for the big precursor tsunami of 1575? Surely this would have left an indelible mark on the landscape of the island? Perhaps it did, but the island was different then and has grown markedly over the centuries as lava flows from the active Mauna Kea volcano have covered the area now occupied by Hilo township. If any evidence survives, it is buried deep beneath these flows, and scientists will have to look to the other islands in their search for earlier events.

After the 1960 tsunami, scientists also learned more about how tsunamis interact with coral reefs and small steep islands. The radio announcements in 1960 of the small wave heights at Tahiti gave a false sense of security to many in Hawaii. But Tahiti is surrounded by coral reefs. Reefs tend to break up tsunami waves, dispersing and absorbing their energy. Furthermore, small, steep islands tend to be minimally affected by tsunami waves—the tsunami simply doesn't "feel" them as it moves through the ocean. However, the tremendous damage done by the waves in Japan was more difficult to explain. Scientists now believe that there were three main reasons why the tsunami was so large and destructive there. As in Hilo Bay, resonance between the tsunami waves and the period of seiche of bays along the Sanriku coast may have played an important role in increasing the size of the waves at the heads of these bays.

A second reason for large waves impacting the coast of Japan has to do with the geographic locations of Chile and Japan at opposite ends of the Pacific Ocean and the shape of the Pacific Ocean basin in between. Imagine a ripple spreading out from the edge of a pond. As it heads toward the middle, the ripple crest is stretched completely across the pond. Then as the ripple approaches the opposite side, the length of the crest is compressed into an increasingly shorter distance. As the length of the crest is stretched, its height decreases; conversely, as the length is compressed, the height increases. As the tsunami waves spread out from Chile across the Pacific Ocean, they were forced to cover an increasingly larger distance until they had reached the middle of the Pacific Ocean basin. But then as they approached Japan, the length of the tsunami crests would cover a progressively shorter distance, resulting in increased wave heights along the coast of Japan.

This being the case, why should there have been such large waves in Hawaii, a small, steep island lying in the middle of the Pacific Ocean? This brings us to the third reason the waves were especially large in Japan. The depths of the Pacific Ocean basin vary enormously from deep abyssal hills and plains at approximately 18,000 ft (5,500 m) to the shallow, young ocean spreading centers such as the East Pacific Rise at depths of only approximately 8,000 ft (2,400 m). Because the speed of tsunami waves is controlled by water depth, the parts of a tsunami crest that pass over the East Pacific Rise, or other shallow features such as the Hawaiian Ridge or the Ontong-Java Plateau, slow down. Meanwhile, the parts of a tsunami crest on either side, still in deeper water, continue at a greater speed. This results in the tsunami wave crest bending toward and around shallow features. The process is called wave refraction, and it can produce a focusing of wave energy, much like light waves can be focused with a magnifying glass to burn a hole in a piece of paper. The shallow features of the Pacific Ocean floor are arranged in such a way that a tsunami generated in Chile will be strongly focused along the coast of Japan and, albeit to a lesser extent, on the Hawaiian Islands. So, the combined effects of resonance in coastal bays, geographic compression, and focusing of wave energy resulted in tsunami waves from Chile causing devastation in both Hawaii and Japan, nearly half a world away.

Half a world away in a different direction is New Zealand. The two most significant tsunamis in New Zealand's history are the 1868 and 1960 events. The 1868 tsunami was generated by a magnitude 8.5 earthquake offshore from Chile's northernmost coast and, like the 1960 tsunami, was a Pacific-wide tsunami. It produced wave heights more than 23 ft (7 m) in New Zealand, the highest ever recorded for a Pacific-wide tsunami in that country. In Hawaii, however, they only reached approximately 13 ft (4 m). As we know, the 1960 tsunami was caused by a massive magnitude 9.5 earthquake in southern Chile, but the wave heights in New Zealand were significantly lower than those of the 1868 event. This was completely the opposite in Hawaii. Why?

Although the 1960 earthquake was much larger than that of 1868, the orientation of the plate boundary and location of the fault rupture where the earthquake occurred meant that the direction of the tsunami's main energy spread across the Pacific Ocean to the north of New Zealand and hence Figure 14.1 starts to make sense. What is perhaps slightly more worrying for New Zealand is that the area where the 1868 earthquake happened has two rather unique qualities. First, geological evidence shows us that every now and again it can produce an earthquake of a magnitude that equals that of the 1960 earthquake. Second, the area is known as a seismic gap, a segment of an active fault known to produce significant earthquakes that has not moved in an unusually long time. The question to be faced, therefore, is whether we will see a repeat of an 1868-type event next time, which will have devastating consequences for a coastline that has seen rapid development in the ensuing century and a half, or whether it will be significantly larger—a 1960 equivalent.

We have been remiss again in our telling of the tsunami story. There are many devastating ocean-wide events that have occurred in the past, and although we have visited both the Pacific and Atlantic Oceans, we have so far avoided the Indian Ocean. How is that possible? It is the one that shocked the modern world, and it is the one that we have left until last—a special place reserved for what was truly a tragedy of global proportions.

15
Boxing Day
The World's Worst Disaster of the 21st Century

> The mixture of fear and fascination.... Most people walked toward not away from the danger... we felt... that we could watch it until it was too close for comfort and then simply step out of its way. Is that how rabbits feel when they stare at oncoming headlights?
>
> **—Kimina Lyall (2006, p. 114)**

The 2004 Indian Ocean tsunami is the largest disaster of the modern era. Not a lot comes close to this. To say that it has been well covered by the media and a seemingly endless stream of books and reports is an understatement. And yet, among all of this there were the people, the survivors, the ones who for whatever reason did not die that day.

These are the stories of the people... and they warrant telling.

The Basics

The 2004 Indian Ocean tsunami was the result of an undersea earthquake that registered a magnitude of approximately 9.1 and caused by a rupture along the fault line demarking the boundary between the Indo-Australian and Eurasian plates, or more precisely the Indian and Burma plates.

This was a big one, and the statistics are morbidly fascinating. On December 26, 2004, approximately 1,000 miles (1,600 km) of fault ruptured, with the seafloor moving around 50 ft (15). It all began when an approximately 62-mile (100-km)-long portion of the plate boundary ruptured and slipped over a period of about 1 minute. This then continued northward, unzipping the seafloor at approximately 6,700 miles per hour (mph; 3 km/second) for 4 minutes, slowing to 5,600 mph (2.5 km/second) for the next 6 minutes. This was the main unzipping phase although it did continue for a little while after that. The earthquake lasted for more than 10 minutes, and it was so large that it caused the entire planet to vibrate by more than 0.5 inches (1 cm).

The energy released by the earthquake was equivalent to more than 1,500 times that of the Hiroshima bomb (a morbid comparison to the tragedy of the first atomic bomb dropped by the United States on August 6, 1945), and the total energy of the tsunami waves was equivalent to approximately 21 petajoules of TNT (5 megatons), which is more than twice the total explosive energy used during all of World War II (including the two atomic bombs).

The tsunami was up to 100 ft (30 m) high and killed an estimated 227,898 people in 14 countries as far west as South Africa on the other side of the Indian Ocean. The Indonesian city of Banda Aceh reported the largest number of victims, with Indonesia as a whole reporting 131,028 dead. The earthquake was one of the deadliest natural disasters in recorded history. With 543 deaths, this tsunami ranks as Sweden's worst natural disaster in terms of the deaths of its nationals.

But what about the tsunami warning system?

The Pacific Tsunami Warning System (PTWS) had been in place in some way, shape, or form since the late 1940s and was well established by the late 1960s following several deadly tsunamis in US waters. By the 2000s, there was also a well-established network of sea-level gauges and deep-sea sensors linked by satellite to round-the-clock monitoring stations based in Hawaii, Alaska, and Japan.

There was no such system in the Indian Ocean.

The quake occurred just before 8:00 a.m. Sumatran time (1:00 a.m. Greenwich Mean Time). Eight minutes later, an alarm was triggered at the Pacific Tsunami Warning Center in Hawaii by seismic signals transmitted from stations in Australia. At 8:14 a.m., an alert notified all countries participating in the network about the quake, indicating that it posed no threat of a tsunami to the Pacific. An hour later, after the size of the earthquake had been revised, a second alert was sent out warning of a possible tsunami in the Indian Ocean. Frantic phone calls were made to issue warnings. But without procedures in place for the Indian Ocean, it was a lottery. They called embassies, the military, local government officials, anybody. At best, the response was disorganized and lethargic. Even the few people who were aware of the dangers were stymied by a lack of preparation, bureaucracy, and inadequate infrastructure. Most, however, either did not know how to interpret the warning signs or were indifferent to them.

Millions of people were at the mercy of the approaching waves.

For many living along coastlines close to the earthquake source, the lack of a warning was a moot point, with the northern coast of Sumatra being struck by a tsunami within 15 minutes. Sri Lanka, India, and Thailand were hit approximately 90 minutes to 2 hours later, and it eventually reached the coast of South Africa, approximately 5,300 mi (8,500 km) away, some 11 hours after the earthquake—two people died there. Although these were the most distant deaths, the tsunami would be recorded in far-flung locations such as Antarctica, Brazil, New Zealand, the United States, and Vanuatu.

Nearly a quarter of a million people died, but many more survived—many by luck, ingenuity, or ancient knowledge. And if surviving the tsunami was not enough, then came the recovery.

Smong

Smong is the term used in the local Devayan language of the people of Simeulue Island to describe "the ocean coming onto the land"—in other words, a tsunami. Simeulue Island sits approximately 60 miles (100 km) off the coast of Aceh Province, Sumatra, Indonesia, and approximately 185 miles (300 km) from the city of Banda Aceh, where more than 100,000 people were killed by the deadly waves of the 2004 Indian Ocean tsunami. Yet on Simeulue, with a population of more than 78,000, only 7 people died as a result of the tsunami, although 35 were killed by the earthquake. Why was there such a huge difference in the tsunami death toll between Banda Aceh and Simeulue? There are a couple of reasons.

First, in 1907 Simeulue had been struck by an earthquake and tsunami, which according to legend killed 70% of the population of the island. Bodies were found in the tops of coconut trees more than 32 ft (10 m) high, and others washed several miles inland. This tsunami became part of the indigenous culture of the island and was transmitted to the population over generations through stories, poems, songs, and lullabies. Cool.

But the story is a bit more complicated.

Aceh Province is Indonesia's most devoutly Muslim region, home to the Free Aceh Movement [Gerakan Aceh Merdeka (GAM)], a rather extremist religious/political group demanding independence and a larger share of Indonesia's oil money. An estimated 15,000 people were killed during this struggle, which lasted from 1976 until 2005. Simulue Island, however, was isolated from the violent insurgency and maintained a local culture of cooperation, which included passing on local indigenous knowledge. The difference in reaction to the Boxing Day earthquake and tsunami is perhaps best illustrated by examining the response of two coastal villages—the village of Jantang on the mainland of Aceh Province and the village of Langi on Simulue.

At Jantang, when people first heard loud noises as the tsunami struck, they thought they were hearing gunshots or explosions, which many suspected were due to nearby fighting between government forces and GAM. Some stayed in the "safety" of their homes, and others avoided heading to high ground because the GAM insurgents were known to hide out in the hills. At Jantang, it was estimated that half of its population of 10,000 was killed by the tsunami.

At Langi on the northern coast of Simulue, the part of the island closest to the earthquake's epicenter, the tsunami arrived only about 8 minutes after the quake, with waves rising 30–50 ft (10–15 m) above sea level. The entire village of Langi was completely destroyed by the tsunami, but of the village's population of approximately 800, not a single person was killed. They had all evacuated to high ground, some even to prearranged meeting places. The power of indigenous knowledge and the need for tsunami education were revealed by the needless tragedy in Aceh.

Some of the personal accounts of this event are harrowing, and unlike many of the older stories presented in this book, these are far closer to the present day. The coastal cities, towns, and villages were modern communities much like they are today, and so we can empathize more intimately with the survivors' experiences.

Life Goes On

Rinaldiana had gotten up in the morning at her house in Ulelheue village, Gampong Pie town, just outside Banda Aceh, and was heading into town when she first felt the earthquake. She said, "First it shake slow, slow . . . but then it shook harder and harder." She was far from home by now, when "suddenly I saw the buildings move . . . the windows are all smashed. I realize then, this is not an ordinary earthquake."

She rushed back home, where she met her husband and two children. Then they saw people running and asked, "Why do you run? And they said. The seawater is coming up." They immediately joined the people running up the road, but it was now not only crowded with people but also jammed with cars; everyone was in a panic. Rinaldiana rushed up the outside stairs of a nearby house and looked toward the ocean and saw "the water was so high like a wave . . . higher than coconut tree." She called down to her husband and oldest daughter, "dear, run, run, run!" but it was too late. As the house began to collapse, she held her youngest daughter in her arms, and suddenly she was hit and lost consciousness. When she came to, her daughter was gone and she was underwater. "I thought it was the end of the world, but I saw the mountains were still there, I saw trees . . . and I realize that this is not doomsday, *I realize that I had to live*, and so I tried to live." Rinaldiana grabbed tree branches and tried to hold on while the water sucked back out to sea. Then she realized that she was on top of a three-story house in another village, 5 miles (8 km) from her own. She looked around and saw dead bodies, then a young boy floated by and he yelled, "Sister, help me . . ." then suddenly he was silent. He had been decapitated by a piece of metal roofing.

Later that morning, Rinaldiana was evacuated to the Harapan Bunda Hospital. In the afternoon, her extended family found her, but there was no sign of her husband and two children. Three months later, she returned to Banda Aceh and continued to search for her lost husband and children, but they were never found. They were lost to the tsunami. Rinaldiana has since remarried with another tsunami survivor, and they have two sons.

Life must go on.

Saved by the Lucky Cushion

Christmas vacation was over and it was time for many of the guests to leave the lovely Golden Buddha Beach Resort on the tiny island of Koh Phra Thong, lying just off the southwest coast of Thailand approximately 430 miles (700 km) across the Andaman Sea from where the giant earthquake had occurred. Some people at the resort had felt mild tremors from the distant quake, whereas other felt nothing. One woman doing her morning meditation noticed ripples in a bowl of water; another thought her husband was gently trying to shake her awake. Others assumed that people were walking on the wooden decks of their beachfront cabins, and one even thought that the Indian military was testing nuclear bombs again.

At approximately 8:30 a.m., two scientists staying at the resort heard a loud noise emanating from the sea and thought it might be dynamite fishing, but they concluded that this was unlikely because this type of fishing had been stopped some years previously.

What no one suspected was a tsunami.

It was now 9.00 a.m., and Chittri Phetcharat, or Do as she was called, was working the reception desk to help customers check out. Suddenly, she heard a very loud noise coming from the beach. At first, she thought it might be an airplane crashing into the sea, and along with other staff and many guests, she ran to the beach. There they saw a "giant wave . . . like none I had ever seen before." She ran back to the office to lock up the money so she could return to the beach and continue to watch this strange phenomenon. Back at the beach, she saw that the water had been sucked out to sea and there was a white line on the horizon—the next, bigger wave. They were all mesmerized watching the waves approach (Figure 15.1-see color plate section).

Suddenly, someone yelled to everybody to rush to the two hills behind the beach, so-called Monkey Mountain and Hornbill Hill (Figure 15.2-see color plate section). Do said, "Honestly, I want to go to the mountain but I look at my friend—no want to go to mountain, so I decide to stay on the beach and watch what is going to be happening." Do stayed with her friend until she saw how close the rushing water was, and then they both started to run. There was no time to get to the hills, so they climbed up a tree. Just then, a big wave hit, knocking them from the tree and her friend disappeared into the swirling sea. Do tried to swim, but she was not a good swimmer and was so scared she wanted to cry but tears just wouldn't come. Then a large cushion popped up next to her and she grabbed hold. The cushion was red, a lucky color—a protective color that symbolizes life force and preservation in Buddhist culture. The wave began sucking Do out to sea on her lucky cushion. She saw other staff members yelling for help and she too screamed out, but no one could hear them over the sound of the tsunami.

After approximately 2 hours, Do saw a couple floating nearby on a sofa and swam toward them, finally reaching them and grabbing the sofa with her right hand. "My left-hand side I hold my cushion that saved my life all this time." The man was English but his wife was Thai, so Do and the other woman began praying together "in Buddhist way." As time passed, they realized that they had to make a decision: to stay with the sofa and possibly float out to sea and be lost or to try to swim to shore. As Do stated, "If we leave the sofa and if wave come, we will die. But if we hold big sofa, we could not get to beach."

They ultimately decided to leave the sofa and swim to the beach. Luck was with them, and they reached the beach before another wave came. Do was helped from the water by the Englishman, and they all began walking toward the two tiny mountains. As they got near Monkey Mountain, some of Do's fellow staff members called to her to join them. She spent one night on Monkey Mountain and learned how the mountain got its name as she was attacked by a crazed monkey during the night. The next day, the injured people on the island were taken to a hospital by helicopter, and Do and others were evacuated to the mainland by speedboat. At the Golden Buddha Beach Resort, four staff members—Duan, a driver; Ae, a waitress; Lumpan, the maintenance foreman; and Lung Luay, a gardener—along with nine guests were killed by the tsunami. But Do survived thanks to her lucky red cushion and good Karma.

Scientists would later estimate that the highest wave to hit Thailand was at Koh Phra Thong. It was nearly 65 ft (20 m) high.

Dance Practice for the New Year's Festival

December 26 was Sunday, a day when school at the Thai village of Ban Thale Nok would normally be closed, and so it was the perfect day for a teacher to help his students practice the dance celebrating the upcoming New Year's festival. Marisa Khunpakdee and six other students had met at their dance teacher's house at 10:00 a.m. to begin cooking lunch when they noticed a white line on the ocean out toward the horizon. Their teacher told them to head down to the beach so that they could watch the strange phenomenon. While they were staring at the sea, a police patrol came by and shouted to them that there had been a big "earthquake and a giant wave at Phuket." Their teacher immediately shouted "run," and they all tried to get into his car but not everyone could fit and so Marisa and her friend, Ismael, began running back toward the teacher's house, when they were picked up by the swirling water. They held on to each other as long as they could, and then "I was floating on the water to the paddy field, then float to the shrimp pond. I saw people, I shout to them to help me. I hold the rock and crawl up." People searching for their relatives came to Marisa's aid and carried her to high ground. From there, she was evacuated by an ambulance.

Of the seven students and their teacher, only the two who hadn't gotten into the car, Marisa and Ismael, would survive. Marisa would soon learn that her mother, sister, grandmother, and two nephews were also killed by the tsunami.

Marisa's school at Ban Thale Nok was in a beautiful but very dangerous location, right next to the ocean. The only thing left of the school following the tsunami was the concrete pad on which it had been built.

Following the tsunami, foreign aid began to flood into Thailand to help rebuild homes and essential infrastructure such as schools. The Thai government was given a grant from Japan to rebuild the Ban Thale Nok school— on the same site next to the sea. We were working in the area collecting data on the tsunami, interviewing survivors, and assisting with preparedness plans for future tsunami events. There was absolutely no way we would quietly go along with plans to rebuild a school on the same site. Imagine the number of students and teachers who would have been killed if the 2004 tsunami had struck on a school day during class hours. However, regional authorities were concerned that they might lose the funds if the school was not "rebuilt" where it had been before.

The struggle over the school site eventually made it all the way to the Crown Prince of Thailand, who agreed that the old site was unsafe and the school should be built on a new site outside the tsunami hazard zone. Not only is the new school safely outside the hazard zone but also it serves as an evacuation center for the village during tsunami warnings. It was a small victory, but every one helps.

The Death Train

Samudra Devi means Queen of the Sea in Sinhala, the official language of the island nation of Sri Lanka. Queen of the Sea was the name given to the train that ran on the coastal rail line from the capital of Sri Lanka at Colombo to the town of Galle at the southern end of the island. It was Sri Lanka's most popular tourist train and on December 26, a holiday weekend for both Buddhists and Christians, it was packed with some 1,500 paying customers and at least 200 other passengers, many of whom were hitching a free ride on top of the train.

The Queen of the Sea left Colombo just before 7.00 a.m., which, due to the time difference, was already well after the massive earthquake off Sumatra had been registered on seismographs in Sri Lanka. The problem was that no one there thought it possible for a tsunami to reach all the way across the Indian Ocean to Sri Lanka. At approximately 9.30 a.m., the train was pulling into the Telwatta Railway Station, located between the towns of Peraliya and Telwatta, when the track was suddenly blocked by tsunami waves surging ashore. The first wave frightened passengers, but many believed they would be safe inside the carriages and so they stayed in their seats. Indeed, many locals climbed on top of the train thinking that they would also be safe from the waves up there. However, the second wave, estimated to be as high as 30 ft (9 m), overtopped the train by some 10 ft (3 m). Two of the carriages were overturned and washed off the tracks, while the others stayed in place but were completely filled with water, drowning nearly all of those still inside. Even those on top of the train did not escape, as most were washed off and crushed by debris. The locomotive was carried 330 ft (100 m) inland, coming to rest in a swamp; both the engineers died at their posts.

Of the estimated 1,700 passengers on the train, only approximately 150 survived, although the actual number killed will never be known because only 900 bodies were recovered. The nearby town of Peraliya also lost hundreds of residents.

This was a classic example of a failure to communicate. There had been an attempt to stop the train earlier at the Ambalangoda Station farther north, but the station crew was so busy dealing with the train that nobody answered the telephone call with the warning.

Following the tsunami, the train cars that were still more or less intact and on the tracks were moved to the Hikkaduwa Railway Station, where they sat on a siding as shown in Figure 15.3, serving as a reminder of the

Figure 15.3 (a) The train at Hikkaduwa. (b) The "new island" of Vilufushi.
Source: W. Dudley.

event. Eventually, a single train car was moved to the Community Tsunami Museum in Peraliya. The locomotive and two train carriages were salvaged and repaired, and a wave was painted on the side of the locomotive, serving as a memorial to those who died. Every year on December 26, the locomotive and two surviving carriages return to Peraliya as part of a religious ceremony to remember those lost to the tsunami.

Train service still operates between Colombo and Galle, but sadly the Queen of the Sea will go down in history as the Death Train. It is the worst train disaster in global railroad history.

The Far Side of India

Even farther away from Sumatra and the origin of the tsunami than Sri Lanka lay Kerala, India. Looking at a map, it appears to be protected by both Sri Lanka and the eastern coast of India, but that would merely give those who lived there a false sense of security. It would turn out to be one of the areas of India most severely hit by the tsunami, as the waves refracted around Sri Lanka to the western side of India.

Five-year-old Haritha lived with her parents and grandparents in Azeekal, a tiny strip of land between the ocean and the Kollan-Kottapuram Waterway. On the morning of the tsunami, she had just returned from the temple with her grandmother to her family's home:

> My mother fed me and put my little sister down to sleep. Mother was washing the clothes, and then suddenly someone said that they saw big waves coming. Stones started falling off the seawall and we began to run. My grandfather carried me and my mother carried my little sister, who was still sleeping. My mother was running and then she fell in a hole. The baby slipped from her hands.

Just as Haritha heard her mother yell, "I lost her," she was washed out of her grandfather's arms. But Haritha would be very lucky: "Someone caught me, carried me, and put me in a boat. I was saved, but we lost my little sister."

Pursothanam, a retired fisherman, also lived in Azeekal. His daughter had come to his house for prayers that morning. After eating, she looked outside and saw the first wave. She told her father that she needed to rush home to her house near the sea in order to move some things inside in case the water rose. He tried to convince her to wait until things calmed down, but she insisted on returning home. "As she stepped out of our house, a dark cloud of water suddenly just gushed in and my daughter was carried away. Her child floated away." He searched for them but never found them. It was not until much later that his daughter's body was recovered; she appeared to have died from being struck by debris.

Mr. Srinivasan, the head government official in the region, explained that there were so many dead bodies that autopsies could not be performed on all of them. The bodies they did examine, however, indicated that the people had died not from drowning but, rather, from blunt force trauma when struck by debris. Even the best swimmers can't survive in an ocean filled with deadly debris. The local government also believed that it needed to cremate and bury the bodies as quickly as possible for fear of disease. This largely unjustified fear had significant consequences, causing increased psychological distress because traditional ceremonies could not be performed. Other more pragmatic issues arose later for the survivors, such as legal problems affecting inheritance and remarriage, and even diplomatic tensions following the cremation of foreign tourists.

The Maldives

Southwest of Sri Lanka and 620 miles (1,000 km) from the coast of India is the Republic of the Maldives, a remote island nation. It consists of 1,192 coral islands but has only 115 square miles (298 km^2) of dry land. The average ground height in the Maldives is less than 8 ft (2.4 m) above sea level, making the Maldives a country whose very existence is threatened by rising sea level. As the Director of the Ministry of Home Affairs stated, "We already had environmental problems, erosion, waves." But prior to December 26, 2004, "We did not know what a tsunami is."

All that would change at 9:30 a.m.

There would be no warning. As Hassan Fulhu Ali, Island Chief of Dhuvaafaru Kaheeb, told us, "It was just a normal day. No one could have predicted it was going to be a fatal day." Many people had just finished their morning prayers when they heard a strange noise. Some said they heard a sound of "bubbling, like rice cooking." As the waves got closer to shore, the noise grew louder—was it a helicopter or a plane? Some, however, knew something unusual was happening. People began to see "the water was springing with bubbles from the ground," and according to 65-year-old Eesa Ali, "The water was boiling." People looked up to see waves coming ashore. Hawwa Aneesa, a street cleaner on the island of Gan, at the southern end of the Maldives, said the first wave "was very small, the second wave bigger, and the third wave very big . . . on top of the palm trees."

Everyone ran for their safety and went to the nearest mosque. The mosques were often the tallest and most solidly built structures on the islands, hence the safest places to wait out the tsunami.

A total of 57 islands suffered serious damage, and 6 were virtually destroyed, with total damage estimated at $400 million. The death toll throughout the Maldives was 102 islanders and 6 visitors. The mosques on high ground no doubt saved many people, as did the fact that fishing is the main source of revenue for the Maldives and many fishermen were at sea. None of them even knew that a tsunami had struck until they returned to their home ports. Furthermore, the country is composed of small, low island atolls, so tsunami waves do not build up to the great heights that they do on continents and large, high islands. The highest wave measured was a modest 14 ft (4.3 m). Although this may seem to be a blessing, it was still almost 6 ft (2.0 m) higher than the average height of the islands.

The island culture played an important role in the recovery. Numerous survivors who had been moved to other islands, either temporarily or in some cases permanently, told us, "We were all treated as their family." But not everything went so well. Many bodies were never recovered, and families seeking closure through burial ceremonies came into conflict over bodies that were recovered, in some cases nearly unrecognizable. Aishath Latheefa of Gan told us about searching for her daughter: "Someone was claiming it was their daughter. So, the police showed me the jewelry and dresses. It was my daughter. The body was bloated, so they had to cut the dress."

As with Thailand, not everything went smoothly with foreign aid in response to the tsunami. Many foreign relief agencies had very little or no knowledge of the local culture and economy. In one particular case, every fisherman in a certain village was given a new fishing boat with an outboard motor. However, once the boats ran out of gas, they were abandoned. The agency didn't realize that the fishermen in the village used boats without motors, instead relying on sails and stern paddles.

Different aid agencies had and still have different ideas and standards for "reconstructing" homes for those displaced. In Thailand, for example, we observed that some agencies had homes built to what would be considered modest standards in Europe or the United States but luxurious by local standards, whereas a neighboring village rebuilt by the local government would basically replace the pre-existing, more humble abodes. This resulted in jealousy and animosity between neighboring villages, the last thing needed in a country following a disaster and already divided by political and religious conflict.

Therefore, it is good to report on one of the best examples of "getting it right" in the Maldives. Vilifushi, north of Gan, was heavily impacted by the

2004 tsunami, with nearly all of the housing and 90% of its infrastructure destroyed. It desperately needed to be rebuilt. Maryam Gasim, her husband, and five children were residents of Vilifushi. They had seen their home and livelihood literary washed away. Gasim stated, "Out of nowhere came the British Red Cross. This is almost like a miracle. Now I will finally be able to get a new home."

The British Red Cross worked closely with the community "to achieve a common consensus because this concerns the very future of our community," said Shiuth Ibrahim, a member of the island's steering committee. Their concern was not only to rebuild critical infrastructure and make the island safe to inhabit but also to treat all the island's citizens fairly. Beginning in 2005, the island was expanded "fourfold" in size and raised an additional 4.6 ft (1.4m) in elevation. A new harbor and new coastal defenses were constructed, with care taken to protect the atoll's coral reef. More than 250 identical three-bedroom homes were built and assigned by drawing in a lottery, but with extended families able to have homes near each other. Funding came from 17 different organizations worldwide, including the United Nations, but the project was managed by the British Red Cross, which got it right. The islanders were finally able to move back to their island in 2009. Vilufushi is now one of the islands that the Maldives government has designated as "a Safe Island in the event of similar natural disasters."

Education

Another lesson we can learn from the 2004 Indian Ocean tsunami was revealed at a beach resort in Thailand. A 10-year-old English schoolgirl named Tilly Smith was on vacation with her parents at Maikhao Beach in northern Phuket. She had recently learned about tsunamis in geography class at school, and when the sea began to withdraw, she recognized that as one of the signs of a tsunami. She alerted her parents, who along with resort staff helped evacuate the beach. It was the only oceanfront resort struck by the tsunami in Thailand where there were no casualties.

The power of education had saved many lives.

Irony

Indonesia was able to establish a warning system using a network of 21 DART (Deep-ocean Assessment and Reporting of Tsunamis) buoys donated by the

United States, Germany, and Malaysia following the 2004 Indian Ocean tsunami. By June 2006, the Indian Ocean Tsunami Warning System was officially established....... just in time for the Pangandaran, Java, earthquake and tsunami that occurred on July 17, 2006.

But despite having a warning system, no warning was given.

Once more, it was left to the PTWC (and the Japan Meteorological Agency) to post a tsunami watch, based on the occurrence of a 7.2-magnitude earthquake centered near the coast of Java. The bulletin was sent out to Indonesia within 30 minutes of the quake, but there was no means to transmit the warning to the people who needed to know—those who lived along the coast. Many of those who felt the earthquake responded by moving away from the shore, but not with any urgency, and not everyone evacuated. Some 668 people would die.

Has the system improved since then? Alas, now the Indian Ocean system has some ongoing problems. It still works but some of the buoys have been damaged by vandals, and others have been stolen by pirates or merely had their batteries ripped off. Not all of the DART buoys are currently operational, and replacements has been delayed due to lack of funds.

It seems we never learn.

The DART buoy system was hugely innovative, and in 2004 the National Oceanic and Atmospheric Administration's (NOAA) Pacific Marine Environmental Laboratory in the United States received the Department of Commerce Gold Medal "for the creation and use of a new moored buoy system to provide accurate and timely warning information on tsunamis" (the prototype is on permanent display at the Pacific Tsunami Museum in Hilo, Hawaii; it was a gift from one of the dedicated scientists, Dr. Frank Gonzalez of the Pacific Marine Environmental Laboratory). But just to prove that we do indeed never learn, in the 2018 budget justification for NOAA, the Trump administration proposed eliminating the DART system altogether as part of a 56% cut to the tsunami warning program.

Maybe when a tsunami strikes Mar-a-Lago in Florida, the administration's opinion will change?

16
Afterword

> There were old traditions of such earthquake waves on this coast, one of two centuries ago doing some damage, and a tsunami of forty years ago and a lesser one of 1892 flooding the streets of Kamaishi and driving people to upper floors and the roofs of their houses. The barometer gave no warning, no indication of any unusual conditions on June 15, and the occurrence of thirteen light earthquake shocks during the day excited no comment. Rain had fallen in the morning and afternoon, and with a temperature of 80° to 90° the damp atmosphere was very oppressive. The villagers on that remote coast adhered to the old calendar in observing their local fetes and holidays, and on that fifth day of the fifth moon had been celebrating the Girls' Festival. Rain had driven them indoors with the darkness, and nearly all were in their houses at eight o'clock, when, with a rumbling as of heavy cannonading out at sea, a roar, and the crash and crackling of timbers, they were suddenly engulfed in the swirling waters. Only a few survivors on all that length of coast saw the advancing wave, one of them telling that the water first receded some 600 yards from ghastly white sands and then the Wave stood like a black wall 80 feet in height, with phosphorescent lights gleaming along its crest. Others, hearing a distant roar, saw a dark shadow seaward and ran to high ground, crying "Tsunami! Tsunami!" Some who ran to the upper stories of their houses for safety were drowned, crushed, or imprisoned there, only a few breaking through the roofs or escaping after the water subsided.
>
> **—Eliza Ruhamah Scidmore (1896)**

The first mention of the word "tsunami" in the English language was in Eliza Scidmore's 1896 article in *National Geographic*. It mentions many of the things we have covered in this book and also raises several points that we haven't.

We always assume that when a tsunami strikes, "we" will be ready. It will probably be a nice sunny afternoon, there will have been a big earthquake, the sirens will sound, and we will walk calmly inland and uphill away from danger.

Sure thing.

Scidmore notes that the 1896 tsunami in Japan happened in the dark. Ah, right, good point, not much we can do about that then. Yes, there is—we can be prepared. They can happen any time and along almost any coastline, fresh or saltwater.

Be prepared.

The 2009 South Pacific tsunami and how things went in American Samoa is one of the best examples of the difference that tsunami education and preparedness training make when a tsunami strikes. During the summer and fall of 2009, there had been extensive tsunami preparedness training in American Samoa for residents and at schools.

According to William Augafa, who lives in the village of Asili near the west end of the island, "most of the old people didn't think it would ever happen. Lucky the chief of the village rang the bell and kids running all over the place saying 'tsunamis coming, tsunamis coming.'" I asked him where the kids learned about tsunamis. He replied, "They learned from school and practice we do here in the villages."

It clearly paid off.

Peter Gurr, an official with the American Samoa Department of Agriculture, told of how school training had saved many lives. As he was driving through a nearby village just before the tsunami struck after the earthquake,

> We noticed there was no kids in front of the school. We found out later because they had training on the Friday before. They had tsunami training . . . and it saved that school . . . around 800–900 kids and . . . they were all in the back of the school heading up the mountain. And at the village of Poloa, the vice principal there saved that school. . . . The cooks were making breakfast that morning and they said let's have breakfast before the kids leave. The young vice principal said, "No we go, we eat later" and as soon as they got to safety the wave hit.

Hundreds of children's lives were saved because they had tsunami evacuation training the week prior to the earthquake and tsunami. They knew what to do and did it.

The key message here is education—consider Tilly Smith in 2004 (see Chapter 15).

Much of the material in this book—the personal stories—is archived at the Pacific Tsunami Museum in Hilo, Hawaii. Without the museum's work, we would never have been exposed to many of the experiences we have written about in this book—and there are many more. This museum is unique in the world, but it drives home a strong message through exhibits, public lectures, tours, and much more, all underpinned by the latest information from around the world. It is also located along one of the most tsunami-exposed coastlines in the world, but it is built to survive. The building in which it is housed has taken everything that Pacific tsunamis can throw at it so far—its physical presence stands as a monument to successfully living with tsunamis. However, as if to underscore the overall global malaise in tsunami education and awareness, it is a nonprofit charitable organization that receives little significant funding from international (or national) governments.

It is true that there are many worthy causes and many other hazards that demand attention. Unfortunately, one of the greatest problems we face in this modern age is the media. They are constantly dragging people on to the next sensational news story; after all, they have to sell news or else they will cease to exist. This has increasingly led to short attention span disasters that, however devastating, seem to drift in and out of social media platforms so rapidly that our senses are overwhelmed by a continual succession of things that are bad for us. But the case for taking tsunamis far more seriously than we do at the moment is a strong one.

This book has shown many ways that tsunamis can be generated and has presented examples of some of the more iconic events. However, there are probably many people out there who are smugly sitting back thinking that their bit of coastline is fine—"it will never happen here." It is invariably at just such moments that something bad does happen, and let us not forget that tsunamis often do not happen by themselves: They are caused by something, and that something can also be devastating. Furthermore, if a tsunami occurs without any warning, it will probably not happen at a convenient time and you will most definitely not be able to simply pick up your life and continue as normal the next day. Consider Japan: Reconstruction work and the rebuilding of lives continue to this day, more than a decade after the devastating 2011 Tōhoku earthquake and tsunami.

And let's consider the "drug" used by almost everyone today—instant communication and being plugged into the world. When we say everyone, it is not just social media but also the media in general, banking, global trade, etc. It is a drug that almost all facets of a "modern" society crave.

In immediate post-2011 Japan, the domestic communications network almost collapsed. People were lining up at public phones, which took priority over cell phone networks. Cell phones were working, but the service was often overloaded and there were far fewer towers in operation. The breaking of multiple submarine cables seriously impacted intra-Asia and trans-Pacific communications and internet access. The cable cuts impacted not only the Japanese carriers but also other Asian carriers. China Telecom reported a decrease of up to 22% of its overall trans-Pacific capacity. One could rationalize this by saying that this was a somewhat unusual event, but then that is what tsunamis are—this is the kind of thing that can happen.

Submarine cables carry the bulk of international e-traffic. For example, approximately 99% of Australia's annual international voice and data traffic worth tens of billions of Australian dollars to the annual e-economy passes along submarine cables. To continue with this example, the December 2006 Taiwan earthquake (no tsunami here, but wait, there's more) cut submarine cables, interrupting telecommunication traffic throughout Southeast Asia and Australasia for approximately 2 months. Although electronic traffic in and out of Australia was slowed, it did not completely cease because more than 70% of all Australian daily traffic volume is routed via Hawaii!

There is a perfect storm brewing, not just for Australia but also for much of e-commerce throughout the world. A decent landslide off one of the Hawaiian Islands that breaks numerous cables, a resultant tsunami that destroys many of the anchor points where cables come ashore—Who repairs it all? If you have an hour or two spare, please wander down this rabbit hole. Submarine cables, tsunamis, and what causes them—they do not mix.

And such perfect storms have happened in the past. The 1929 Grand Banks earthquake, submarine landslide, and tsunami near Newfoundland, Canada, was just such a beast. The magnitude 7.2 earthquake triggered a large submarine landslide that snapped 12 submarine transatlantic telegraph cables and produced a tsunami up to 26 ft (8 m) high that killed approximately 28 people. All means of communication were cut off by the destruction, plus it happened in the dark, plus relief and recovery efforts were hampered by a blizzard that struck the day after.

As if such perfect storms are not enough, we have not even mentioned meteorological tsunamis or meteotsunamis. They are a book in themselves, but having eased our minds with the knowledge that tsunamis are usually generated by earthquakes, landslides, or some type of volcanic event—albeit with the occasional human-induced one or an asteroid or something—these come out of left field.

A 2015 article by Šepić and colleagues began by stating,

> A series of tsunami-like waves of non-seismic origin struck several southern European countries during the period of 23 to 27 June 2014. The event caused considerable damage from Spain to Ukraine. Here, we show that these waves were long-period ocean oscillations known as meteorological tsunamis which

> are generated by intense small-scale air pressure disturbances. . . . This is the first documented case of a chain of destructive meteorological tsunamis occurring over a distance of thousands of kilometres. (p. 1)

Again, if you spend a little time researching these, you will soon find out that officially these are large waves that are driven by air-pressure disturbances often associated with fast-moving weather events, such as severe thunderstorms, squalls, and other storm fronts. The storm generates a wave that moves toward the shore, and it is amplified by a shallow continental shelf and inlet, bay or other coastal feature. Unfortunately, like so much to do with tsunamis, there is much that we still do not know about the magnitude, frequency, or even locations where these have happened. However, we do know that they can easily create waves up to 6 ft (2 m) high and can occur in many places throughout the world, including the Great Lakes of North America, the Gulf of Mexico, the Atlantic coast, and the Mediterranean and Adriatic Seas.

That ticks off a few more coastlines that we have not mentioned previously, and just to show that tsunamis yet again are a worryingly dangerous hazard that can be generated by so many different processes, meteotsunamis may well increase in the future as rising temperatures due to climate change are likely to be associated with increased storminess.

In closing, it is important to remind readers that there are things we can do, there are things being done, and science moves on. Everyone has an obligation to look after themselves and also those less capable of doing so. Be prepared: If you live at or near any coast, simply know what you need to do to get to safety. If you go on vacation to some beautiful exotic location, be prepared in the same way. It is not a big deal; it is just like knowing to look for oncoming traffic when you want to cross the road—it is simply something you do. Relying on the "authorities" or others to tell you what to do is a cop-out. They may or may not warn you. Of course, tsunami warning systems are continuously being improved (subject to funding), and there is always some education and awareness activity going on, but let's be honest—our efforts as educators/scientists/whatever are a drop in the ocean.

Please be aware.

Please help others.

Tsunamis have happened in the past, and they will happen in the future.

> Those that fail to learn from history, are doomed to repeat it.
>
> **—Winston Churchill (House of Commons, November 16, 1948)**

Bibliography

Introduction

Goff, J., Chagué-Goff, C. (2014). The Australian tsunami database—A review. *Progress in Physical Geography 38*, 218–240.

Goff, J., Terry, J., Chagué-Goff, C., Goto, K. (2014). What is a megatsunami? *Marine Geology 358*, 12–17. DOI: 10.1016/j.margeo.2014.03.013

National Oceanic and Atmospheric Administration, Center for Tsunami Research. (2020). *DART® (Deep-ocean Assessment and Reporting of Tsunamis)*. Retrieved from https://nctr.pmel.noaa.gov/Dart

Chapter 1. The Case of the Disappearing Lighthouse

Fryer, G. J., Watts, P., Pratson, L. F. (2004). Source of the great tsunami of 1 April 1946: A landslide in the upper Aleutian forearc. *Marine Geology 203*, 201–218.

Kanamori, H. (1972). Mechanism of tsunami earthquakes. *Physics of the Earth and Planetary Interiors 6*, 346–359.

López, A. M, Okal, E. A. (2006). A seismological reassessment of the source of the 1946 Aleutian "tsunami" earthquake. *Geophysical Journal International 165*, 835–849.

Macdonald, G., Shepard, F., Cox, D. (1947). The tsunami of April 1, 1946, in the Hawaiian Islands. *Pacific Science 1*, 21–37.

Okal, E. A., Synolakis, C. E., Fryer, G. J., Heinrich, P., Borrero, J. C., Ruscher, C., Arcas, D., Guille, G., Rousseau, D. (2002). A field survey of the 1946 Aleutian tsunami in the far field. *Seismological Research Letters 73*, 490–503.

Rutherford, D. (1986). Disaster at Scotch Cap. *The Keeper's Log 2*, 12–14.

Shepard, F., Macdonald, G., Cox, D. (1950). The tsunami of April 1, 1946. 6. *Bulletin of the Scripps Institution of Oceanography of the University of California*, 391–528.

US Geological Survey. (n.d.). *M-8.6—1946 Aleutian Islands (Unimak Island) earthquake, Alaska*. Retrieved from https://earthquake.usgs.gov/earthquakes/eventpage/official19460401122901_30/executive

USBeacons.com. (n.d.). *Scotch cap light*. Retrieved from https://www.usbeacons.com/lt.cgi?lighthouse=Scotch+Cap+Light.

Transcript Sources

Herbert Nishimoto Collection, Pacific Tsunami Museum.

Marsue McGinnis Collection, Pacific Tsunami Museum.

Chapter 2. How Weird Squiggles Led from Sheaves of Rice to the Depth of the Seas

Atwater, B. F., Musumi-Rokkaku, S., Satake, K., Tsuji, Y., Ueda, K., Yamaguchi, D. K. (2016). *The orphan tsunami of 1700: Japanese clues to a parent earthquake in North America*. Seattle, WA: University of Washington Press.

Data.gov. (2020). *Archival and discovery of December 23, 1854 tsunami event on marigrams.* Retrieved from https://catalog.data.gov/dataset/archival-and-discovery-of-december-23-1854-tsunami-event-on-marigrams

Kauahikaua, J., Poland, M. (2012). One hundred years of volcano monitoring in Hawaii. *Eos 93*, 29–30.

Kitamura, A., Kobayashi, K. (2014). Geologic evidence for prehistoric tsunamis and coseismic uplift during the AD 1854 Ansei–Tokai earthquake in Holocene sediments on the Shimizu Plain, central Japan. *The Holocene 24*, 814–827.

National Oceanic and Atmospheric Administration. (2019). *Science on a sphere. Tsunami historical series: Aleutian Islands—1946.* Retrieved February 2020 from https://sos.noaa.gov/datasets/tsunami-historical-series-aleutian-islands-1946

Theberge, A. E. (2005). *150 years of tides on the western coast: The longest series of tidal observations in the Americas.* Washington, DC: National Oceanic and Atmospheric Administration.

Tsuchiya, Y., Shuto, N. (Eds.). (1995). *Tsunami: Progress in prediction, disaster prevention and warning* (Vol. 4). New York, NY: Springer.

Whitmore, P. M. (2009). Tsunami warning systems. In E. Bernard (Ed.), *The sea. Volume 15: Tsunamis* (pp. 401–442). Cambridge, MA: Harvard University Press,.

Chapter 3. Voices from the Past

Clark, S. K. (2010). A shift in scientific literacy: Earthquakes generate tsunamis. *Eos 91*, 316–317.

Goff, J. (in press). The stuttering Polynesian diaspora—Environmental crises of the past and an uneasy future. In D. Sugawara, K. Yamada (Eds.), *Island civilizations: Implications to the future of the Earth.* Dordrecht, the Netherlands: Springer.

Goff, J., Chagué-Goff, C. (2015). Three large tsunamis on the non-subduction, western side of New Zealand over the past 700 years. *Marine Geology 363*, 243–260. doi:10.1016/j.margeo.2015.03.002

Goff, J., Goto, K., Ebina, Y., Terry, J. (2016). Defining tsunamis: Yoda strikes back? *Earth-Science Reviews 159*, 271–274. doi:10.1016/j.earscirev.2016.06.003

Goff, J., McFadgen, B. G., Chagué-Goff, C., Nichol, S. L. (2012). Palaeotsunamis and their influence on Polynesian settlement. *The Holocene 22*, 1061–1063. doi:10.1177/0959683612437873Goto, K., Chagué-Goff, C., Fujino, S., Goff, J., Jaffe, B., Nishimura, Y., Richmond, B., Suguwara, D., Szczuciński, W., Tappin, D. R., Witter, R., Yulianto, E. (2011). New insights into tsunami risk from the 2011 Tohoku-oki event. *Marine Geology 290*, 46–50. doi:10.1016/j.margeo.2011.10.004

Grace, A. (1907). *Folktales of the Maori.* Wellington, New Zealand: Gordon & Gotch.

King, D., Goff, J. (2010). Benefitting from differences in knowledge, practice and belief: Māori oral traditions and natural hazards science. *Natural Hazards and Earth System Sciences 10*, 1927–1940. doi:10.5194/nhess-10-1-2010

King, D. N., Shaw, W. S., Meihana, P., Goff, J. (2018). Māori oral histories and the impact of tsunamis in Aotearoa-New Zealand. *Natural Hazards and Earth System Sciences 18*, 907–919. Retrieved from https://doi.org/10.5194/nhess-18-907-2018

Macdonald, G., Shepard, F., Cox, D. (1947). The tsunami of April 1, 1946, in the Hawaiian Islands. *Pacific Science 1*, 21–37.

Meanyu, E. S. (1923). *Origin of Washington geographic names.* Seattle, WA: University of Washington Press.

Muhlhausler, P. (2014). Reducing Pacific languages to writings. In: J. E. Joseph, T. J. Taylor (Eds.), *Ideologies of language. Routledge Library Editions, Linguistics A: General Linguistics* (Vol. 4, pp. 189–205). London, UK: Routledge.

Sakata, M. (2011). Possibilities of reality, variety of versions: The historical consciousness of Ainu folktales. *Oral Tradition 26*, 175–190.

Shepard, F., Macdonald, G., Cox, D. (1950). The tsunami of April 1, 1946. *Bulletin of the Scripps Institution of Oceanography of the University of California 6*, 391–528.

von Haast, J. (1877). Address. *Transactions and Proceedings of the New Zealand Institute 10*, 37–56.

Watt, V. (Ed.). (2004). *The Cambridge dictionary of English place-names: Based on the collections of the English Place-Name Society*. Cambridge, UK: Cambridge University Press.

Chapter 4. The World's Oldest Tsunami Victim at the Gateway to the Pacific—and Beyond

Anderson, A., Chappell, J., Gagan, M., Grove, R. (2006). Prehistoric maritime migration in the Pacific islands: An hypothesis of ENSO forcing. *The Holocene 16*, 1–6.

Bedford, S., Spriggs, M., Buckley, H., Valentin, F., Regenvanu, R. (2009). The Teouma Lapita site, South Efate, Vanuatu: A summary of three field seasons (2004–2006). In P. Sheppard, T. Thomas, G. Summerhayes (Eds.), *Lapita: Ancestors and descendants* (New Zealand Archaeological Association Monograph Series, pp. 215–234). Auckland, New Zealand: New Zealand Archaeological Association.

Bedford, S., Spriggs, M., Regenvanu, R. (2006). The Teouma Lapita site and the early human settlement of the Pacific Islands. *Antiquity 80*, 812–828.

Davies, H. L. (2002). Tsunamis and the coastal communities of Papua New Guinea. In J. Grattan, R. Torrence (Eds.), *Natural disasters and cultural change* (pp. 28–42). London, UK: Routledge.

Davies, H. L. (2016). *Aitape story: The great New Guinea tsunami of 1998*. Sydney, Australia: Halstead.

Durband, A. C., Creel, J. A. (2011). A reanalysis of the early Holocene frontal bone from Aitape, New Guinea. *Archaeology in Oceania 46*, 1–5.

Fenner, F. J. (1941). Fossil human skull fragments of probable Pleistocene age from Aitape, New Guinea. *Records of the South Australian Museum 6*, 335–356.

Goff, J., Golitko, M., Cochrane, E., Curnoe, D., Terrell, J. (2017). Reassessing the environmental context of the Aitape Skull—The oldest tsunami victim in the world? *PLoS One 12*(10), e0185248. Retrieved from https://doi.org/10.1371/journal.pone.0185248

Goff, J., McFadgen, B. G., Chagué-Goff, C., Nichol, S. L. (2012). Palaeotsunamis and their influence on Polynesian settlement. *The Holocene 22*, 1061–1063. doi:10.1177/0959683612437873

Golitko, M., Cochrane, E. E., Schechter, E. M., Kariwiga, J. (2014). Archaeological and palaeoenvironmental investigations near Aitape, northern Papua New Guinea. *Journal of Pacific Archaeology 7*, 139–150.

Hossfeld, P. S. (1949). The stratigraphy of the Aitape skull and its significance. *Transactions of the Royal Society of South Australia 72*, 201–207.

Hossfeld, P. S. (1964). The Aitape calvarium. *Australian Journal of Science 27*, 179.

Hossfeld, P. S. (1965). Radiocarbon data and palaeoecology of the Aitape fossil human remains. *Proceedings of the Royal Society of Victoria 76*, 161–165.

McSaveney, M., Goff, J., Darby, D., Goldsmith, P., Barnett, A., Elliott, S., Nongkas, M. (2000). The 17th July 1998 tsunami, Sissano Lagoon, Papua New Guinea—Evidence and initial interpretation. *Marine Geology 170*, 81–92.

Mulrooney, M. A., Bickler, S. H., Allen, M. S., Ladefoged, T. N. (2011). High-precision dating of colonization and settlement in East Polynesia: A comment on Wilmshurst et al. *Proceedings of the National Academy of Sciences of the USA 108*, E192–E194.

Nason-Jones, J. (1930). *The geology of the Finsch coast area*. London, UK: Harrison & Sons.

Phillips, B., Neal, D., Wikle, T., Subanthore, A., Hyrapiet, S. (2008). Mass fatality management after the Indian Ocean tsunami. *Disaster Prevention and Management 17*, 681–697.

Roullier, C., Benoit, L., McKey, D. B., Lebot, V. (2013). Historical collections reveal patterns of diffusion of sweet potato in Oceania obscured by modern plant movements and recombination. *Proceedings of the National Academy of Sciences of the USA 110*, 2205–2210.

Specht, J., Denham, T., Goff, J., Terrell, J. E. (2014). Deconstructing the Lapita cultural complex in the Bismarck Archipelago. *Journal of Archaeological Research 22*, 89–140. doi:10.1007/s10814-013-9070-4

Storey, A. A., Ramirez, J. M., Quiroz, D., Burley, D. V., Addison, D. J., Walter, R., Anderson, A. J., Hunt, T. L., Athens, J. S., Huynen, L., Matisoo-Smith, E. A. (2007). Radiocarbon and DNA evidence for a pre-Columbian introduction of Polynesian chickens to Chile. *Proceedings of the National Academy of Sciences of the USA 104*, 10335–10339.

Terrell, J. E., Pope, K. O., Goff, J. (2011). Context and relevance. In J. E. Terrell, E. M. Schechter (Eds.), *Archaeological investigations on the Sepik coast, Papua New Guinea* (Fieldiana Anthropology series, Vol. 42, pp. 21–28). Chicago, IL: Field Museum of Natural History.

Wilmshurst, J. M., Hunt, T. L., Lipo, C. P., Anderson, A. J. (2011). High-precision radiocarbon dating shows recent and rapid initial human colonization of East Polynesia. *Proceedings of the National Academy of Sciences of the USA 108*, 1815–182.

Chapter 5. The Monster of Lituya Bay

Dall, W. H. (1883). *United States Coast and Geodetic Survey Pacific Coast Pilot*, Part I, p. 203.

Emmons, G. T. (1911). Native Account of the Meeting between La Perouse and the Tlingit. *Anthropologist New Series 13*, 294–298.

Fritz, H. M., Mohammed, F., Yoo, J. (2009). Lituya Bay landslide impact generated mega-tsunami 50th anniversary. *Pure & Applied Geophysics 166*, 153–175. doi:10.1007/s00024-008-0435-4

Mertie, J. B. (1933). *Notes on the geography and geology of Lituya Bay*. Washington, DC: US Government Printing Office.

Miller, D. J. (1960a). *Giant waves in Lituya Bay, Alaska* (US Geological Survey Professional Paper 354-C, pp. 51–86). Reston, VA: US Geological Survey.

Miller, D. J. (1960b). The Alaska earthquake of July 10, 1958: Giant wave in Lituya Bay. *Bulletin of the Seismological Society of America 50*, 253–266.

Chapter 6. The Perils of Freshwater Tsunamis

Anonymous. (1931a, March 5). Huge slips in Napier District. *Evening Post CXI*(54), p. 11. Retrieved from http://paperspast.natlib.govt.nz/cgi-bin/paperspast?a=d&cl=search&d=EP19310305.2.48&srpos=2&e=-------10--1----0waikare+landslide

Anonymous. (1931b, March 19). Seen from the air. Damage on the East Coast. Napier-Wairoa shoreline. Aftermath of earthquake. *Auckland Star LXII*(66), p. 8. Retrieved from https://paperspast.natlib.govt.nz/newspapers/AS19310319.2.76?end_date=19-03-1931&query=seen+from+the+air&snippet=true&start_date=19-03-1931&title=AS

Bonney, T. G. (1868). *The alpine regions of Switzerland and the neighbouring countries: A pedestrian's notes on their physical features, scenery, and natural history*. London, UK: Deighton, Bell & Company.

Bosa, S., Petti, M. (2013). A numerical model of the wave that overtopped the Vajont dam in 1963. *Water Resources Management 27*, 1763–1779.

Chagué-Goff, C., Nichol, S. L., Jenkinson, A. V., Heijnis, H. (2000). Signatures of natural catastrophic events and anthropogenic impact in an estuarine environment, New Zealand. *Marine Geology 167*, 285–301.

de Lange, W. P., Moon, V. G. (2016). Volcanic generation of tsunamis: Two New Zealand palaeoevents. In G. Lamarche, J. Mountjoy, S. Bull, T. Hubble, S. Krastel, E. Lane, A. Micallef, L. Moscardelli, C. Mueller, I. Pecher, S. Woelz (Eds.), *Submarine mass movements and their consequences* (Advances in Natural and Technological Hazards Research, Vol. 41, pp. 559–567). Cham, Switzerland: Springer.

Donaldson, G., Goff, J., Chagué, C., Gadd, P., Fierro, D. (2019). The Waikari River tsunami: New Zealand's largest historical event. *Sedimentary Geology 383*, 148–158. Retrieved from https://doi.org/10.1016/j.sedgeo.2019.02.006

Fritz, H. M., Mohammed, F., Yoo, J. (2009). Lituya Bay landslide impact generated megatsunami 50th anniversary. *Pure & Applied Geophysics 166*, 153–175. doi:10.1007/s00024-008-0435-4

Grégoire de Tours. (563). Histoire des Francs. Livres 1 à 6, page de frontispice. Luxeuil ou Corbie, fin du VII e siècle. BnF, Manuscrits, Latin 17655 fol. 2.

Harbitz, C. B., Glimsdal, S., Løvholt, F., Kveldsvik, V., Pedersen, G. K., Jensen, A. (2014). Rockslide tsunamis in complex fjords: From an unstable rock slope at Åkerneset to tsunami risk in western Norway. *Coastal Engineering 88*, 101–122.

Higman, B., Shugar, D. H., Stark, C. P., Ekström, G., Koppes, M. N., Lynett, P., Dufresne, A., Haeussler, P. J., Geertsema, M., Gulick, S., Mattox, A. (2018). The 2015 landslide and tsunami in Taan Fiord, Alaska. *Scientific Reports 8*, 1–12.

Hull, A. G. (1986). Pre-AD 1931 tectonic subsidence of Ahuriri Lagoon, Napier, Hawke's Bay, New Zealand. *New Zealand Journal of Geology and Geophysics 29*, 75–82.

Kilburn, C. R., Petley, D. N. (2003). Forecasting giant, catastrophic slope collapse: Lessons from Vajont, northern Italy. *Geomorphology 54*, 21–32.

Kremer, K., Marillier, F., Hilbe, M., Simpson, G., Dupuy, D., Yrro, B. J., Rachoud-Schneider, A. M., Corboud, P., Bellwald, B., Wildi, W., Girardclos, S. (2014). Lake dwellers occupation gap in Lake Geneva (France–Switzerland) possibly explained by an earthquake–mass movement–tsunami event during Early Bronze Age. *Earth and Planetary Science Letters 385*, 28–39.

Moore, J. G., Schweickert, R. A., Robinson, J. E., Lahren, M. M., Kitts, C. A. (2006). Tsunami-generated boulder ridges in Lake Tahoe, California–Nevada. *Geology 34*, 965–968.

Rouwet, D., Sandri, L., Marzocchi, W., Gottsmann, J., Selva, J., Tonini, R., Papale, P. (2014). Recognizing and tracking volcanic hazards related to non-magmatic unrest: A review. *Journal of Applied Volcanology 3*, 1–17.

von Hardenberg, W. G. (2011). *Expecting disaster: The 1963 landslide of the Vajont Dam*. Environment & Society Portal, Arcadia, no. 8. Retrieved from http://www.environmentandsociety.org/node/3401

Woodhouse. (1940). *Tales of Pioneer Women*.

Chapter 7. Tsunamis and the US Navy

1868 tsunami destroys a town. (2005, March 25). *Washington Times*. Retrieved from https://www.washingtontimes.com/news/2005/mar/25/20050325-082538-6765r

Borrero, J. C., Lynett, P. J., Kalligeris, N. (2015). Tsunami currents in ports. *Philosophical Transactions of the Royal Society A: Mathematical, Physical and Engineering Sciences 373*(2053), 20140372.

Goff, J. (2012). Tsunamis and stranded vessels: Up Ship Creek without a paddle? *Geographical Research 50*, 102–107. doi:10.1111/j.1745-5871.2011.00705.x

Hall, D. A. (1917a). The wreck of the *U.S.S. De Soto*. *Proceedings of the United States Naval Institute 43*, 1151–1160.

Hall, D. A. (1917b). The launching of the *U.S.S. Monongahela*. *Proceedings of the United States Naval Institute 43*, 1161–1172.

Lander, J. F., Whiteside, L. S., Lockridge, P. A. (2002). A brief history of tsunamis in the Caribbean Sea. *Science of Tsunami Hazards 20*, 57–94.

Shaw, E. (2013). *Operation Unified Assistance: 2004 Sumatran earthquake and tsunami humanitarian relief*. Newport, RI: United States Naval War College, Joint Military Operations Department.

Shaw, R., Rodriguez, H., Wachtendorf, T., Kendra, J., Trainor, J. (2006). A snapshot of the 2004 Indian Ocean tsunami: Societal impacts and consequences. *Disaster Prevention and Management 15*, 163–177.

Ships Nostalgia. (2020). *Port of Arica 1868 HMS Charnasilla*. Retrieved from https://www.shipsnostalgia.com/gallery/showphoto.php/photo/1072226/cat/523/title/port-of-arica-1868-hms-charnasilla

Taussig, J. K. (1926). The tidal wave and earthquake at Arica, Peru, in 1868. *United States Naval Institute Proceedings*, 52/7/281.

Zahibo, N., Pelinovsky, E., Yalciner, A., Kurkin, A., Koselkov, A., Zaitsev, A. (2003). The 1867 Virgin Island tsunami: Observations and modeling. *Oceanologica Acta 26*, 609–621.

Chapter 8. Not Such a Good Friday for Alaska

Baker, F. (2018). *The 1964 earthquake story I'm finally allowed to tell*. Retrieved from https://www.echoak.com/2018/03/1964-earthquake-story

Brocher, T. M., Fuis, G. S., Filson, J. R., Haeussler, P. J., Holzer, T. L., Plafker, G., Blair, J. L. (2014). *The 1964 great Alaska earthquake and tsunamis: A modern perspective and enduring legacies*. Reston, VA: Earthquake Science Center, US Geological Survey. Retrieved from http://pubs.usgs.gov/fs/2014/3018

National Research Council, Committee on the Alaska Earthquake (Ed.). (1972). *The great Alaska earthquake of 1964* (Vol. 1, Pt. 1). Washington, DC: National Academy of Sciences.

US Geological Survey. (2020). *M9.2 Alaska earthquake and tsunami of March 27, 1964*. Retrieved from https://earthquake.usgs.gov/earthquakes/events/alaska1964

Weller, J. M. (1972). Human response to tsunami warnings. In National Research Council, Committee on the Alaska Earthquake (Ed.), *The great Alaska earthquake of 1964: Human ecology* (pp. 222–227). Washington, DC: National Academy of Sciences.

Wood, F. J., Leipold, L. E. (Eds.). (1966). *The Prince William Sound, Alaska, earthquake of 1964 and aftershocks: Research studies: seismology and marine geology. Pt. A. Engineering seismology. Pt. B. Seismology. Pt. C. Marine geology* (Vol. 10, No. 3). Washington, DC: US Government Printing Office.

Interviews by Walter Dudley and Jeanne Johnston for the National Oceanic and Atmospheric Administration, on File at the Pacific Tsunami Museum in Hilo, Hawaii:

• Thelma Barnum and Fred Christoffersen, Valdez, Alaska
• Robert and Mack Eads, Seward, Alaska• Linda McRae McSwain, Seward, Alaska
• Kenneth Lester, Kodiak, Alaska
• Gene Andersen, Ouzinkie, Spruce Island, Alaska

Chapter 9. Strange, But True

Alexander Turnbull Library. (2020). Waimangu geyser: The world's largest Retrieved from https://teara.govt.nz/en/photograph/6497/waimangu-geyser-the-worlds-largest

Anonymous. (1908, November 3). Miramar relics. *Evening Post 76*(108), 4.

Assier-Rzadkiewicz, S., Heinrich, P., Sabatier, P. C., Savoye, B., Bourillet, J. F. (2000). Numerical modeling of a landslide-generated tsunami: The 1979 Nice event. *Pure & Applied Geophysics 157*, 1707–1727.

Cain, G., Goff, J., McFadgen, B. G. (2019). Prehistoric mass burials: Did death come in waves? *Journal of Archaeological Method and Theory 26*, 714–754.

Goff, J., Chagué-Goff, C. (2009). Cetaceans and tsunamis—Whatever remains, however improbable, must be the truth? *Natural Hazards and Earth System Science 9*, 855–857.

Greenberg, D. A., Murty, T. S., Ruffman, A. (1993). A numerical model for the Halifax Harbor tsunami due to the 1917 explosion. *Marine Geodesy 16*, 153–167.

Ioualalen, M., Migeon, S., Sardoux, O. (2010). Landslide tsunami vulnerability in the Ligurian Sea: Case study of the 1979 October 16 Nice international airport submarine landslide and of identified geological mass failures. *Geophysical Journal International 181*, 724–740.

Kernaghan, L., Foot, R. (2017). *Halifax explosion*. Retrieved from https://www.thecanadianencyclopedia.ca/en/article/halifax-explosion

Kremer, K., Anselmetti, F., Evers, F. M., Girardclos, S., Goff, J., Nigg, V., Talkova, E. (in press). Freshwater (paleo)tsunamis – A review. *Earth-Science Reviews*.

Kulikov, E. A., Rabinovich, A. B., Thomson, R. E., Bornhold, B. D. (1996). The landslide tsunami of November 3, 1994, Skagway Harbor, Alaska. *Journal of Geophysical Research: Oceans 101*(C3), 6609–6615.

Leech, T. D. J. (1950). *Project Seal: The generation of waves by means of explosives*. Wellington, New Zealand: Department of Scientific and Industrial Research.

McLeod, H. N. (1912). Pre-Pakeha occupation of Wellington District. *Early Settlers & Historical Association Wellington Journal 1*, 14–17.

Sohn, E. (2020). *Why the Great Molasses Flood was so deadly*. Retrieved from https://www.history.com/news/great-molasses-flood-science

Williams, C. (2020). *Remembering the "Great Gorbals Whisky Flood" which sent a tidal wave of alcohol onto the streets*. Retrieved from https://www.glasgowlive.co.uk/news/history/a-tidal-wave-of-alcohol-17667249

Chapter 10. Megasharknado

Berger, W. H. (2011). Geologist at sea: Aspects of ocean history. *Annual Review of Marine Science 3*, 1–34.

Chapman, C. R., Morrison, D. (1994). Impacts on the Earth by asteroids and comets: Assessing the hazard. *Nature 367*(6458), 33–40.

Dingle, R. V. (1977). The anatomy of a large submarine slump on a sheared continental margin (SE Africa). *Journal of the Geological Society 134*, 293–310.

Goff, J., Chagué-Goff, C., Archer, M., Dominey-Howes, D., Turney, C. (2012). The Eltanin asteroid impact: Possible South Pacific palaeomegatsunami footprint and potential implications for the Pliocene–Pleistocene transition. *Journal of Quaternary Science 27*, 660–670.

Iwatani, H., Irizuki, T., Hayashi, H. (2012). Global cooling in marine climates and local tectonic events in southwest Japan at the Plio–Pleistocene boundary. *Palaeogeography, Palaeoclimatology, Palaeoecology 350*, 1–18.

Ketten, D. R. (1995). Estimates of blast injury and acoustic trauma zones for marine mammals from underwater explosions. In R. A. Kastelein, J. A. Thomas, P. E. Nachtigall (Eds.), *Sensory systems of aquatic mammals* (pp. 391–407). Woerden, the Netherlands: De Spil.

Ma, P., Hou, Q., Chai, C. (1998). Distribution of iridium in different size fractions at the N/Q boundary in loess: Implications for an asteroid impact 2.4 Ma ago. *Acta Scientiarum Naturalium-Universitatis, Pekinensis 34*, 775–782.

Molina, E. (2015). Evidence and causes of the main extinction events in the Paleogene based on extinction and survival patterns of foraminifera. *Earth-Science Reviews 140*, 166–181.

Morrissey, M., Gisler, G., Weaver, R., Gittings, M. (2010). Numerical model of crater lake eruptions. *Bulletin of Volcanology 72*, 1169–1178.

Peng, H. (1994). An extraterrestrial event at the tertiary–quaternary boundary. In *New developments regarding the KT event and other catastrophes in Earth history* (Contribution 825, p. 88). Houston, TX: Lunar and Planetary Institute.

Pierazzo, E., Artemieva, N. (2012). Local and global environmental effects of impacts on Earth. *Elements 8*, 55–60.

Pierazzo, E., Garcia, R. R., Kinnison, D. E., Marsh, D. R., Lee-Taylor, J., Crutzen, P. J. (2010). Ozone perturbation from medium-size asteroid impacts in the ocean. *Earth and Planetary Science Letters 299*, 263–272.

Pimiento, C., Clements, C. F. (2014). When did Carcharocles megalodon become extinct? A new analysis of the fossil record. *PLoS One 9*(10), e111086.

Pimiento, C., Griffin, J. N., Clements, C. F., Silvestro, D., Varela, S., Uhen, M. D., Jaramillo, C. (2017). The Pliocene marine megafauna extinction and its impact on functional diversity. *Nature Ecology & Evolution 1*, 1100–1106.

Pyenson, N. D., Gutstein, C. S., Parham, J. F., Le Roux, J. P., Chavarría, C. C., Little, H., Metallo, A., Rossi, V., Valenzuela-Toro, A. M., Velez-Juarbe, J., Santelli, C. M. (2014). Repeated mass strandings of Miocene marine mammals from Atacama region of Chile point to sudden death at sea. *Proceedings of the Royal Society B: Biological Sciences 281*(1781), 20133316.

Rampino, M. R., Haggerty, B. M. (1996). Impact crises and mass extinctions: A working hypothesis. In G. Ryder, D. E. Fastovsky, S. Gartner (Eds.), *The Cretaceous–Tertiary event and other catastrophes in Earth history* (Issue 307, p. 11). Boulder, CO: Geological Society of America.

Why whales strand. (n.d.). Retrieved June 29, 2020, from https://deafwhale.com

Toon, O. B., Zahnle, K., Morrison, D., Turco, R. P., Covey, C. (1997). Environmental perturbations caused by the impacts of asteroids and comets. *Reviews of Geophysics 35*, 41–78.

Tungsheng, D. Z. L. (1989). Progresses of loess research in China. Part 1: Loess stratigraphy. *Quaternary Sciences.*

Valenzuela-Toro, A. M., Gutstein, C. S., Varas-Malca, R. M., Suarez, M. E., Pyenson, N. D. (2013). Pinniped turnover in the South Pacific Ocean: New evidence from the Plio-Pleistocene of the Atacama Desert, Chile. *Journal of Vertebrate Paleontology 33*, 216–223.

Vermeij, G. J. (2004). Ecological avalanches and the two kinds of extinction. *Evolutionary Ecology Research 6*, 315–337.

Walsh, S. A., Hume, J. P. (2001). A new Neogene marine avian assemblage from north-central Chile. *Journal of Vertebrate Paleontology 21*, 484–491.

Walsh, S. A., Martill, D. M. (2006). A possible earthquake-triggered mega-boulder slide in a Chilean Mio-Pliocene marine sequence: Evidence for rapid uplift and bonebed genesis. *Journal of the Geological Society 163*, 697–705.

Wang, S., Ouyang, Z., Xiao, Z. (1999). Palaeoclimatic cycles, global environmental changes and new glacial periods induced by the impact of extraterrestrial bodies. *Chinese Journal of Geochemistry 18*, 298–304.

Wang, S., Ziyaun, O., Zhifeng, X., Chunlai, L. (1999). Formation of palaeoclimatic cycles and new glacial period induced by impact of extraterrestrial bodies. *Earth Science-Journal of China University of Geosciences 24*(6), 568–572.

Weiss, R., Lynett, P., Wünnemann, K. (2015). The Eltanin impact and its tsunami along the coast of South America: Insights for potential deposits. *Earth and Planetary Science Letters 409*, 175–181.

Yunli, L. U. O., Jiaxin, X. Z. C. (2000). Geologic events around 2.50 MaBP shown in Fengyi Section in Lishi. *Marine Geology & Quaternary Geology 2*, 16.

Zhisheng, A. (1985). Study on lower boundary of Quaternary in North China—An important climatic-geological event in Early Matuyama Chron. *Marine Geology & Quaternary Geology*, 4.

Ziyuan, O., Shijie, W., Zhifeng, X., Lei, Z., Chunlei, L., Wenzhu, L. (1995). Palaeoenvironmental and palaeoclimate catastrophe induced by Cenozoic bolide impact events. *Quaternary Sciences*, 4.

Chapter 11. Saved by the Baguette

Accary, F., Roger, J. (2010). Tsunami catalog and vulnerability of Martinique (Lesser Antilles, France). *Science of Tsunami Hazards 29*, 148–174.

Bruins, H. J., MacGillivray, J. A., Synolakis, C. E., Benjamini, C., Keller, J., Kisch, H. J., Klügel, A., Van Der Plicht, J. (2008). Geoarchaeological tsunami deposits at Palaikastro (Crete) and the Late Minoan IA eruption of Santorini. *Journal of Archaeological Science 35*, 191–212.

Bruins, H. J., van der Plicht, J. (2014). The Thera olive branch, Akrotiri (Thera) and Palaikastro (Crete): Comparing radiocarbon results of the Santorini eruption. *Antiquity 88*, 282–287.

Driessen, J. (2019a). *Collateral damage: The Santorini eruption and the collapse of Minoan civilization*. Retrieved from https://dial.uclouvain.be/pr/boreal/object/boreal:219676

Driessen, J. (2019b). The Santorini eruption: An archaeological investigation of its distal impacts on Minoan Crete. *Quaternary International 499*, 195–204.

Druitt, T. H., McCoy, F. W., Vougioukalakis, G. E. (2019). The Late Bronze Age eruption of Santorini volcano and its impact on the ancient Mediterranean world. *Elements 15*, 185–190.

Francis, Peter, Self, Stephen. (1983). The eruption of krakatau. *Scientific American 249*, 172–187.

Goff, J., Terry, J. (2016). Tsunamigenic slope failures: The Pacific Islands "blind spot"? *Landslides 13*, 1535–1543.

Goodman-Tchernov, B. N., Dey, H. W., Reinhardt, E. G., McCoy, F., Mart, Y. (2009). Tsunami waves generated by the Santorini eruption reached eastern Mediterranean shores. *Geology 37*, 943–946.

Gusiakov, V. K. (2009). Tsunami history: Recorded. In E. N. Bernard, A. R. Robinson (Eds.), *The Sea. Volume 15: Tsunamis* (pp. 23–53). Cambridge, MA: Harvard University Press.

Hasegawa, S. (2017). *Spatial analysis of Māori settlements in New Zealand and its implication for future tsunami management*. Unpublished report, Department of Anthropology, Brown University, Providence, RI.

History's first and only eyewitness account. (2020). Retrieved from http://eruptionmtvesuvius.weebly.com/plinys-letters.html

Jones, W. H. S. (1918). *Pausanias: Description of Greece. Volume I, Books I and II (Attica, Corinth)* (Loeb Classical Library). New York, NY: Putnam's Sons.

Little, L. M., Papadopoulos, J. K. (1998). A social outcast in early Iron Age Athens. *Hesperia 67*, 375–404.

O'Loughlin, K. F., Lander, J. F. (2003). Caribbean tsunamis: A 500-year history from 1498–1998 (Advances in Natural and Technological Hazards Research Vol. 20) Dordrecht, the Netherlands: Kluwer.

Poisson, B., Pedreros, R. (2010). Numerical modelling of historical landslide-generated tsunamis in the French Lesser Antilles. *Natural Hazards and Earth System Sciences 10*, 1281–1292.

Rosen, J. (2015). *Benchmarks: May 8, 1902: The deadly eruption of Mount Pelée*. Retrieved from https://www.earthmagazine.org/article/benchmarks-may-8-1902-deadly-eruption-mount-pelee

Self, S. (1992). Krakatau revisited: The course of events and interpretation of the 1883 eruption. *GeoJournal 28*, 109–121.

Self, S. (2006). The effects and consequences of very large explosive volcanic eruptions. *Philosophical Transactions of the Royal Society A: Mathematical, Physical and Engineering Sciences 364*, 2073–2097.

Self, S., Rampino, M. R. (1981). The 1883 eruption of Krakatau. *Nature 294*, 699–704.

Simkin, T. (1983). *Krakatau, 1883: The volcanic eruption and its effects*. Washington, DC: Smithsonian Institution Press.

Tanguy, J. C. (1994). The 1902–1905 eruptions of Montagne Pelée, Martinique: Anatomy and retrospection. *Journal of Volcanology and Geothermal Research 60*, 87–107.

Thornton, I. W. (1997). *Krakatau: The destruction and reassembly of an island ecosystem*. Cambridge, MA: Harvard University Press.

Vitaliano, D. B. (2007). *Geomythology: Geological origins of myths and legends* (Special Publication No. 273, pp. 1–7). London, UK: Geological Society.

Whipps, H. (2008). *How the eruption of Thera changed the world*. Retrieved from https://www.livescience.com/4846-eruption-thera-changed-world.html

Chapter 12. 1755, Lisbon: The Benefit of Brothels

Accary, F., Roger, J. (2010). Tsunami catalog and vulnerability of Martinique (Lesser Antilles, France). *Science of Tsunami Hazards 29*, 148–174.

Atwater, B. F., Uri, S., Buckley, M., Halley, R. S., Jaffe, B. E., López-Venegas, A. M., Reinhardt, E. G., Tuttle, M. P., Watt, S., Wei, Y. (2012). Geomorphic and stratigraphic evidence for an unusual tsunami or storm a few centuries ago at Anegada, British Virgin Islands. *Natural Hazards 63*, 51–84.

Banerjee, D., Murray, A. S., Foster, I. D. L. (2001). Scilly Isles, UK: Optical dating of a possible tsunami deposit from the 1755 Lisbon earthquake. *Quaternary Science Reviews 20*, 715–718.

Baptista, M. A., Miranda, J. M. (2009). Revision of the Portuguese catalog of tsunamis. *Natural Hazards & Earth System Sciences 9*, 25–42.

Barkan, R., ten Brink, U. S., Lin, J. (2009). Far field tsunami simulations of the 1755 Lisbon earthquake: Implications for tsunami hazard to the U.S. East Coast and the Caribbean. *Marine Geology 264*,109–122. doi:10.1016/j.margeo.2008.10.010

Borlase, W. (1758). *The natural history of Cornwall*. Oxford, UK.

de Almeida Marques, J. O. (2005). The paths of providence: Voltaire and Rousseau on the Lisbon earthquake. *Cadernos de História e Filosofia da Ciência 15*(Series 3), 33–57.

Farrell, E. J., Ellis, J. T., Hickey, K. R. (2015). Tsunami case studies. In *Coastal and marine hazards, risks, and disasters* (pp. 93–128). New York, NY: Elsevier. Retrieved from http://dx.doi.org/10.1016/B978-0-12-396483-0.00004-2

Grossman, M. (2011, May). Joseph Stepling and the Tabor meteorite fall. *Meteorite*, 15–21.

Joel, L. (2020). *Benchmarks: November 1, 1755: Earthquake destroys Lisbon*. Retrieved from https://www.earthmagazine.org/article/benchmarks-november-1-1755-earthquake-destroys-lisbon

Long, D. (2015). *A catalogue of tsunamis reported in the UK*. British Geological Survey, Energy and Marine Geosciences Programme, Internal Report IR/15/043.

Muir-Wood, R., Mignan, A. (2009). A phenomenological reconstruction of the Mw9 November 1st 1755 Lisbon earthquake. In L. A. Mendes-Victor., C. S. Oliveira, J. Azebedo, A. Ribeiro (Eds.), *The 1755 Lisbon earthquake revisited* (pp. 121–146). Dordrecht, the Netherlands: Springer.

Mullin, J. R. (1992). The reconstruction of Lisbon following the earthquake of 1755: A study in despotic planning. *Planning Perspective 7*, 157–179.

Pereira, A. S. (2009). The opportunity of a disaster: The economic impact of the 1755 Lisbon earthquake. *Journal of Economic History 69*, 466–499.

Roger, J., Baptista, M. A., Sahal, A., Accary, F., Allgeyer, S., Hébert, H. (2011). The transoceanic 1755 Lisbon tsunami in Martinique. *Pure & Applied Geophysics 168*, 1015–1031.

Winthrop, J. (1755). *A Lecture on Earthquakes: Read in the Chapel of Harvard-College in Cambridge, NE November 26th 1755. On Occasion of the Great Earthquake which Shook New-England the Week Before*. Edes & Gill, at their printing-office next to the prison in Queen-Street.

Chapter 13. Storegga: No Referendum for This Brexit

Ballin, T. B. (2017). Rising waters and processes of diversification and unification in material culture: The flooding of Doggerland and its effect on north-west European prehistoric populations between ca. 13 000 and 1500 cal BC. *Journal of Quaternary Science 32*, 329–339.

Chakoumakos, B. C. (2004). Preface to the Clathrate Hydrates special issue. *American Mineralogist 89*, 1153–1154.

Cotterill, C. J., Phillips, E., James, L., Forsberg, C. F., Tjelta, T. I., Carter, G., Dove, D. (2017). The evolution of the Dogger Bank, North Sea: A complex history of terrestrial, glacial and marine environmental change. *Quaternary Science Reviews 171*, 136–153.

Dingle, R. V. (1977). The anatomy of a large submarine slump on a sheared continental margin (SE Africa). *Journal of the Geological Society 134*, 293–310.

Evans, H. M., Godwin, M. E., Moir, J. R., Burkitt, M. C. (1932). East Anglian Notes. *Proceedings of the Prehistoric Society of East Anglia 7*, 131–133.

Gaffney, V., Fitch, S., Bates, M., Ware, R. L., Kinnaird, T., Gearey, B., Hill, T., Telford, R., Batt, C., Stern, B., Whittaker, J. (2020). *Multi-proxy evidence for the impact of the Storegga Slide tsunami on the early Holocene landscapes of the southern North Sea*. Retrieved from https://www.biorxiv.org/content/10.1101/2020.02.24.962605v1

Gaffney, V. L., Fitch, S., Smith, D. N. (2009). *Europe's lost world: The rediscovery of Doggerland* (Vol. 160). York, UK: Council for British Archaeology.

Hill, J., Avdis, A., Mouradian, S., Collins, G., Piggott, M. (2017). *Was Doggerland catastrophically flooded by the Mesolithic Storegga tsunami?* Retrieved from https://arxiv.org/abs/1707.05593

Hill, J., Collins, G. S., Avdis, A., Kramer, S. C., Piggott, M. D. (2014). How does multiscale modelling and inclusion of realistic palaeobathymetry affect numerical simulation of the Storegga Slide tsunami? *Ocean Modelling 83*, 11–25.

Kim, J., Løvholt, F., Issler, D., Forsberg, C. F. (2019). Landslide material control on tsunami genesis—The Storegga Slide and tsunami (8,100 years BP). *Journal of Geophysical Research: Oceans 124*, 3607–3627.

King, W. (1863). An attempt to correlate the glacial and post-glacial deposits of the British Isles, and to determine their order of succession. *The Geologist 6*, 168–178.

Long, D., Smith, D. E., Dawson, A. G. (1989). A Holocene tsunami deposit in eastern Scotland. *Journal of Quaternary Science 4*, 61–66.

Waddington, C., Wicks, K. (2017). Resilience or wipe out? Evaluating the convergent impacts of the 8.2 ka event and Storegga tsunami on the Mesolithic of northeast Britain. *Journal of Archaeological Science: Reports 14*, 692–714.

Wagner, B., Bennike, O., Klug, M., Cremer, H. (2007). First indication of Storegga tsunami deposits from east Greenland. *Journal of Quaternary Science 22*, 321–325.

Wallis, J. (1701). A letter of Dr John Wallis, D. D. Professor of Geometry in the University of Oxford, and Fellow of the Royal Society in London; to Dr Hans Sloane, Secretary to the said Royal Society; relating to that isthmus, or neck of land, which is supposed to have joyned England and France in former times, where now is the passage between Dover and Calais. *Philosophical Transactions of the Royal Society of London 22*, 967–979.

Warren, G. (2020). Climate change and hunter gatherers in Ireland: Problems, potentials and pressing research questions. *Proceedings of the Royal Irish Academy: Archaeology, Culture, History, Literature 120C*, 1–22.

Weninger, B., Schulting, R., Bradtmöller, M., Clare, L., Collard, M., Edinborough, K., Hilpert, J., Jöris, O., Niekus, M., Rohling, E. J., Wagner, B. (2008). The catastrophic final flooding of Doggerland by the Storegga Slide tsunami. *Documenta Praehistorica 35*, 1–24.

Chapter 14. 1960 Chile: Did the Earth Move for You?

Acton, J. M., Hibbs, M. (2012). *Why Fukushima was preventable*. Washington, DC: Carnegie Endowment for International Peace.

Camfield, F. E. (1994). Tsunami effects on coastal structures. *Journal of Coastal Research 12*, 177–187.

Chagué, C., Sugawara, D., Goto, K., Goff, J., Dudley, W., Gadd, P. (2018). Geological evidence for the 1946 and 1960 tsunamis in Shinmachi, Hilo, Hawaii? *Sedimentary Geology 364*, 319–333. Retrieved from https://doi.org/10.1016/j.sedgeo.2017.09.010

Cifuentes, I. L. (1989). The 1960 Chilean earthquakes. *Journal of Geophysical Research: Solid Earth 94*(B1), 665–680.

Cisternas, M., Atwater, B. F., Torrejón, F., Sawai, Y., Machuca, G., Lagos, M., Eipert, A., Youlton, C., Salgado, I., Kamataki, T., Shishikura, M. (2005). Predecessors of the giant 1960 Chile earthquake. *Nature 437*(7057), 404–407.

Domínguez, L. (1961). *Diario de la Isla de Pascua 1960–1961*.

Duke, C. M. (1960). The Chilean Earthquakes of May 1960. *Science* 132(3442), 1797–1802.

Fotografías Isla de Pascua: 04-Ahu Tongariki despues del maremoto. (2020). Retrieved from http://lorenzodominguez.com/ARCHIVO/ISLA/04-Ahu%20Tongariki%20despues%20del%20maremoto/index.html

Goff, J., Nichol, S. L., Chagué-Goff, C., Horrocks, M., McFadgen, B., Cisternas, M. (2010). Predecessor to New Zealand's largest historic trans-South Pacific tsunami of 1868 AD. *Marine Geology 275*, 155–165.

Imagina Rapa Nui Easter Island. (2020). *Ahu Tongariki, the 15 moai statues*. Retrieved from https://imaginaisladepascua.com/en/easter-island-sightseeing/easter-island-archaeology/ahu-tongariki

International Atomic Energy Agency. (2011). *IAEA international fact finding expert mission of the Fukushima Dai-Ichi NPP accident following the Great East Japan Earthquake and tsunami*. Report of the IAEA member states. Tokyo, Fukushima Dai-ichi NPP, Fukushima Daini NPP, and Tokai Dai-ni NPP, Japan 24 May–2 June 2011.

Japan Meteorological Agency. (1963). *The report on the tsunami of the Chilean earthquake, 1960*. Technical report of the Japan Meteorological Agency, No. 26. Tokyo, Japan: Japan Meteorological Agency.

Johnston, J. B. (2003). *Personal accounts from survivors of the Hilo tsunamis of 1946 and 1960: Toward a disaster communication model*. Master's thesis, University of Hawaii–Manoa, Manoa, HI.

Kronmüller, E., Atallah, D. G., Gutiérrez, I., Guerrero, P., Gedda, M. (2017). Exploring indigenous perspectives of an environmental disaster: Culture and place as interrelated resources for remembrance of the 1960 mega-earthquake in Chile. *International Journal of Disaster Risk Reduction 23*, 238–247.

León, T., Vargas, G., Salazar, D., Goff, J., Guendón, J. L., Andrade, P., Alvarez, G. (2019). Recording large Holocene paleotsunamis along the hyperarid coastal Atacama Desert in the major northern Chile seismic gap. *Quaternary Science Reviews 220*, 335–358.

Margalef, O., Álvarez-Gómez, J. A., Pla-Rabes, S., Cañellas-Boltà, N., Rull, V., Sáez, A., Geyer, A., Peñuelas, J., Sardans, J., Giralt, S. (2018). Revisiting the role of high-energy Pacific events in the environmental and cultural history of Easter Island (Rapa Nui). *The Geographical Journal 184*, 310–322.

Sievers, C. H. A., Villegas, C. G., Barros, G. (1963). The seismic sea wave of 22 May 1960 along the Chilean coast (translated by P. Saint-Amand). *Bulletin of the Seismological Society of America 53*, 1125–1190.

Taniguchi, T., Woo, D. T. (1961). The seismic wave casualties in Hilo, Hawaii. *Archives of Environment Health 2*, 434–439.

Vitosek, M. J. (1963). The tsunami of 22 May 1960 in French Polynesia. *Bulletin of the Seismological Society of America 53*, 1229–1236.

Western States Seismic Policy Council. (2019). *1960 Chile tsunami*. Retrieved from https://www.wsspc.org/resources-reports/tsunami-center/significant-tsunami-events/1960-chile-tsunami

Chapter 15. Boxing Day—The World's Worst Disaster of the 21st Century

Beardsley, K., McQuinn, B. (2009). Rebel groups as predatory organizations: The political effects of the 2004 tsunami in Indonesia and Sri Lanka. *Journal of Conflict Resolution 53*, 624–645.

Cochard, R., Ranamukhaarachchi, S. L., Shivakoti, G. P., Shipin, O. V., Edwards, P. J., Seeland, K. T. (2008). The 2004 tsunami in Aceh and southern Thailand: A review on coastal ecosystems, wave hazards and vulnerability. *Perspectives in Plant Ecology, Evolution and Systematics 10*, 3–40.

Folger, T. (2018). *Will Indonesia be ready for the next tsunami?* Retrieved from https://www.nationalgeographic.com/news/2018/9/141226-tsunami-indonesia-catastrophe-banda-aceh-warning-science

Geist, E. L., Titov, V. V., Arcas, D., Pollitz, F. F., Bilek, S. L. (2007). Implications of the 26 December 2004 Sumatra–Andaman earthquake on tsunami forecast and assessment models for great subduction-zone earthquakes. *Bulletin of the Seismological Society of America 97*, S249–S270.

Goff, J., Liu, P. L., Higman, B., Morton, R., Jaffe, B. E., Fernando, H., Lynett, P., Fritz, H., Synolakis, C., Fernando, S. (2006). Sri Lanka field survey after the December 2004 Indian Ocean tsunami. *Earthquake Spectra 22*, 155–172.

Jankaew, K., Atwater, B. F., Sawai, Y., Choowong, M., Charoentitirat, T., Martin, M. E., Prendergast, A. (2008). Medieval forewarning of the 2004 Indian Ocean tsunami in Thailand. *Nature 455*, 1228–1231.

Leone, F., Lavigne, F., Paris, R., Denain, J. C., Vinet, F. (2011). A spatial analysis of the December 26th, 2004 tsunami-induced damages: Lessons learned for a better risk assessment integrating buildings vulnerability. *Applied Geography 31*, 363–375.

Lyall, K. (2006). *Out of the blue: Facing the tsunami*. Sydney, Australia: ABC Books.

McAdoo, B. G., Dengler, L., Prasetya, G., Titov, V. (2006). Smong: How an oral history saved thousands on Indonesia's Simeulue Island during the December 2004 and March 2005 tsunamis. *Earthquake Spectra 22*, 661–669.

Morton, R. A., Goff, J. R., Nichol, S. L. (2008). Hydrodynamic implications of textural trends in sand deposits of the 2004 tsunami in Sri Lanka. *Sedimentary Geology 207*, 56–64.

Patten, D. M. (2008). Does the market value corporate philanthropy? Evidence from the response to the 2004 tsunami relief effort. *Journal of Business Ethics 81*, 599–607.

Rodgers, L., Fletcher, G. (2014). *Indian Ocean tsunami: Then and now*. Retrieved from https://www.bbc.co.uk/news/world-asia-30034501

Roos, D. (2018). *The 2004 tsunami wiped away towns with "mind-boggling" destruction*. Retrieved from https://www.history.com/news/deadliest-tsunami-2004-indian-ocean

Samarajiva, R. (2005). Policy commentary: Mobilizing information and communications technologies for effective disaster warning: Lessons from the 2004 tsunami. *New Media & Society 7*, 731–747.

Suppasri, A., Goto, K., Muhari, A., Ranasinghe, P., Riyaz, M., Affan, M., Mas, E., Yasuda, M., Imamura, F. (2015). A decade after the 2004 Indian Ocean tsunami: The progress in disaster preparedness and future challenges in Indonesia, Sri Lanka, Thailand and the Maldives. *Pure and Applied Geophysics 172*, 3313–3341.

Titov, V., Rabinovich, A. B., Mofjeld, H. O., Thomson, R. E., González, F. I. (2005). The global reach of the 26 December 2004 Sumatra tsunami. *Science 3*, 2045–2048.

Uckay, I., Sax, H., Harbarth, S., Bernard, L., Pittet, D. (2008). Multi-resistant infections in repatriated patients after natural disasters: Lessons learned from the 2004 tsunami for hospital infection control. *Journal of Hospital Infection 68*, 1–8.

US Department of Commerce. (2020). *National Data Buoy Center*. Retrieved from https://www.ndbc.noaa.gov/dart.shtml

Chapter 16. Afterword

Bent, A. L. (1995). A complex double–couple source mechanism for the Ms 7.2 1929 Grand Banks earthquake. *Bulletin of the Seismological Society of America 85*, 1003–1020.

Clarke, J. E. H., Shor, A. N., Piper, D. J., Mayer, L. A. (1990). Large-scale current-induced erosion and deposition in the path of the 1929 Grand Banks turbidity current. *Sedimentology 37*, 613–629.

Long, D. (2018). Cataloguing tsunami events in the UK. *Geological Society, London, Special Publications 456*, 143–165.

Moore, A. L., McAdoo, B. G., Ruffman, A. (2007). Landward fining from multiple sources in a sand sheet deposited by the 1929 Grand Banks tsunami, Newfoundland. *Sedimentary Geology 200*, 336–346.

National Oceanic and Atmospheric Administration. (2020). *What is a meteotsunami?* Retrieved from https://oceanservice.noaa.gov/facts/meteotsunami.html

Qui, W. (2011). *Submarine cables cut after magnitude-9.0 earthquake and tsunami in Japan.* Retrieved from https://www.submarinenetworks.com/news/cables-cut-after-magnitude-89-earthquake-in-japan

National Geographic. (2011). *The recent earthquake wave on the coast of Japan.* Retrieved from https://ngslis.wordpress.com/2011/03/18/the-recent-earthquake-wave-on-the-coast-of-japan/

Šepić, J., Vilibić, I., Rabinovich, A. B., Monserrat, S. (2015). Widespread tsunami-like waves of 23–27 June in the Mediterranean and Black Seas generated by high-altitude atmospheric forcing. *Scientific Reports* 5, 1–8.

Index

For the benefit of digital users, indexed terms that span two pages (e.g., 52–53) may, on occasion, appear on only one of those pages.